Digital Interface Handbook

Digital Interface Handbook

Third edition

Francis Rumsey

and

John Watkinson

 Routledge
Taylor & Francis Group

LONDON AND NEW YORK

First published 1993
Second edition 1995
Third edition 2004

This edition published 2013 by Focal Press

Published 2017 by Routledge
2 Park Square, Milton Park, Abingdon, Oxon OX14 4RN
711 Third Avenue, New York, NY 10017, USA

First issued in hardback 2017

Routledge is an imprint of the Taylor & Francis Group, an informa business

British Library Cataloguing in Publication Data
Rumsey, Francis
   Digital interface handbook. – 3rd ed.
      1.   Interface circuits – Handbooks, manuals, etc.   2.   Digital electronics –
      Handbooks, manuals etc.
      I.   Title   II.   Watkinson, John, 1950 –
      621.3'897

Library of Congress Cataloguing in Publication Data
A catalogue record for this book is available from the Library of Congress

ISBN 13: 978-1-138-40829-6 (hbk)
ISBN 13: 978-0-240-51909-8 (pbk)

Typeset by Newgen Imaging Systems (P) Ltd, Chennai, India

Contents

1

Introduction to interfacing

1.1 The need for digital interfaces

1.1.1 Transparent links

Digital audio and video systems make it possible for the user to maintain a high and consistent sound or picture quality from beginning to end of a production. Unlike analog systems, the quality of the signal in a digital system need not be affected by the normal processes of recording, transmission or transferral over interconnects, but this is only true provided that the signal remains in the digital domain throughout the signal chain. Converting the signal to and from the analog domain at any point has the effect of introducing additional noise and distortion, which will appear as audible or visual artefacts in the programme material. Herein lies the reason for adopting digital interfaces when transferring signals between digital devices – it is the means of ensuring that the signal is carried 'transparently', without the need to introduce a stage of analog conversion. It enables the receiving device to make a 'cloned' copy of the original data, which may be identical numerically and temporally.

1.1.2 The need for standards

The digital interface between two or more devices in an audio or video system is the point at which data is transferred. Digital interconnects allow programme data to be exchanged, and they may also provide a certain capacity for additional information such as 'housekeeping data' (to inform a receiver of the characteristics of the programme signal, for example), text data, subcode data replayed from a tape or disk, user data, communications channels (e.g. low quality speech) and perhaps timecode. These applications are all covered in detail in the course of this book. Standards have been developed in an attempt to ensure that the format of the data adheres to a convention and that the meaning of different bits is clear, in order that devices may communicate correctly, but there is more than one standard and thus not all devices will communicate with each other. Furthermore, even between devices using ostensibly the same interface there are often problems in communication due to differences in the level or completeness of implementation of the standard, the effects of which are numerous. Older devices may have problems with data from newer devices or vice versa, since the standard may have been modified or clarified over the years.

As digital audio and video systems become more mature the importance of correct communication also increases, and the evidence is that manufacturers are now beginning to take correct implementation of interface standards more seriously, since more people are adopting fully digital signal chains. But it must be said that at the same time the applications of such technology are becoming increasingly complicated, and the additional data which accompanies programme data on many interfaces is becoming more comprehensive – there being a wide range of different uses for such data. Thus manufacturers must decide the extent to which they implement optional features, and what to do with the data which is not required or understood by a particular device. Eventually it is likely that digital interface receivers will become more 'intelligent', such that they may analyse the incoming data and adapt so as to accommodate it with the minimum of problems, but this is rare in today's systems and would currently add considerably to the cost of them.

1.1.3 Digital interfaces and programme quality

To say that signal quality cannot be affected provided that the signal remains in the digital domain is a bold statement and requires some qualification, since there will be cases where signal processing in the digital chain may affect quality. The statement is true if it is possible to assume that the sampling rate of the signal, its resolution (number of bits per sample) and the method of quantization all remain unchanged (see Chapter 2). Further one must assume that the signal has not been subjected to any processing, since filtering, gain changing, and other such operations may introduce audible or visual side effects. Operations such as sampling frequency conversion and changes in resolution (say, between 20 and 16 bits per sample) may also introduce artefacts, since these can never be 'perfect' processes. Therefore an operation such as copying a signal digitally between two recording systems with the same characteristics is a transparent process, resulting in absolutely no loss of quality (see Figure 1.1), but copying between two recorders with different sampling rates via a sampling frequency convertor is not (although the side effects in most cases are very small).

Confusion arises when users witness a change in sound or picture quality even in the former of the above two cases, leading them to suggest that digital copying is *not* a transparent process, but the root of this problem is not in the digital copying process – it is in the digital-to-analog conversion process of the device which the operator is using to monitor the signal, as discussed in greater detail in Chapter 2. It is true that timing instabilities and (occasionally) errors may arise when signals are transferred digitally between devices, but both of these are normally correctable or avoidable within the digital domain. Since data errors are extremely rare in digitally interfaced systems, it is timing instabilities which will have the most likely effect on the convertor. Poor quality clock recovery in the receiver and lack of timebase correction in the convertor often allow timing instabilities resulting from the digital interface to affect programme quality when it is monitored in the analog domain. This does *not* mean that the digital programme itself is of poor quality, simply that the convertor is incapable of rejecting the instability. Although it is difficult to ensure low jitter clock recovery in receivers, especially with the stability required for very high convertor resolutions

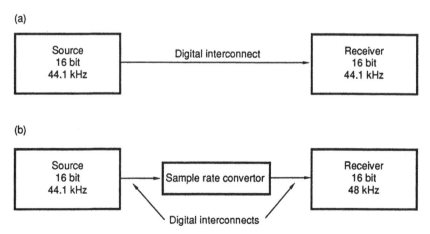

Figure 1.1   (a) A 'clone' copy may be made using a digital interconnect between two devices operating at the same sampling rate and resolution. (b) When sampling parameters differ, digital interconnects may still be used, such as in this example, but the copy will not be a true 'clone' of the original.

(e.g. 20 bits in audio), it is definitely here that the root of the problem lies and not really in the nature of digital signals themselves, since it is possible to correct such instabilities with digital signals but not normally possible with analog signals. This is discussed further in Chapter 2 and in section 6.4.3, and for additional coverage of these topics the reader is referred to Rumsey[1] and Watkinson[2,3].

1.2 Analog and digital communication compared

Before going on to examine specific digital audio and video interfaces it would be useful briefly to compare analog and digital interfaces in general, and then to look at the basic principles of digital communication.

In an analog wire link between two devices the baseband audio or video signal from a transmitter is carried directly in the form of variations in electrical voltage. Any unwanted signals induced in the wire, such as radio frequency interference (RFI) or mains hum will be indistinguishable from the wanted signal at the receiver, as will any noise or timing instability introduced between transmitter and receiver, and these are likely to affect signal quality (see Figure 1.2). Techniques such as balancing (see section 1.7.1) and the use of transmission lines (see section 1.7.3) are used in analog communication to minimize the effects of long lines and interference. Forms of modulation are used in analog links, especially in radio frequency transmission, whereby the baseband (unmodulated) signal is used to alter the characteristics of a high frequency carrier, resulting in a spectrum with a sideband structure around the carrier. Modulation may give the signal increased immunity to certain types of interference[4], but modulated analog signals are rarely carried over wire links within a studio installation.

Pulse amplitude modulation (PAM) is a means of modulating an analog signal onto a series of pulses, such that the amplitude of the pulses varies according to the instantaneous amplitude of the analog signal (see Figure 1.3) and this is the basis of pulse code modulation (PCM) whereby the amplitude of PAM pulses is quantized, resulting in a binary signal that is

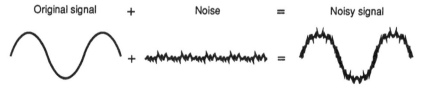

Original signal   +   Noise   =   Noisy signal

Figure 1.2 If noise is added to an analog signal the result is a noisy signal. In other words, the noise has become a feature of the signal which may not be separated from it.

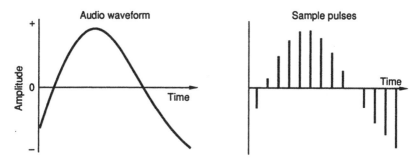

Figure 1.3 In pulse amplitude modulation (PAM) a regular chain of pulses is amplitude modulated by the baseband signal.

exceptionally immune to noise and interference since it only has two states. This process normally takes place in the analog-to-digital (A/D) convertor of an audio or video system, described in more detail in Chapter 2. PCM is the basis of all current digital audio and video interfaces, and, as shown in Figure 1.4, the effects of unwanted noise and timing instability may be rejected by reclocking the binary signal and comparing it with a fixed threshold. Provided that the receiving device is able to distinguish between the two states, one and zero, and can determine the timing slot in which each binary digit (bit) resides, the system may be shown to be able to reject any adverse effects of the link. Over long links a digital signal may be reconstructed regularly, in order to prevent it from becoming impossibly distorted and difficult to decode. Of course there will be cases in which an interfering signal will cause a sufficiently large effect on the digital signal to prevent it from being correctly reconstructed, but below the threshold at which this happens there will be no effect at all.

Compared with an analog baseband signal, its digital counterpart normally requires a considerably greater bandwidth, as can be seen from the diagrams above, but the advantage of a digital signal is that it can normally survive over a channel with a relatively low signal-to-noise ratio. Another advantage of digital communications is that a number of signals of different types may be carried over the same physical link without interfering with each other, since they may be time-division multiplexed (TDM), as discussed in section 1.5.5.

1.3 Quantization, binary data and word length

When a PAM pulse is quantized it is converted into a PCM word of a certain length or resolution. The number of bits in the word determines

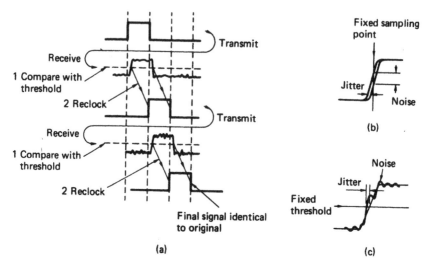

Figure 1.4 (a) A binary signal is compared with a threshold and reclocked on receipt, thus the meaning will be unchanged. (b) Jitter on a signal can appear as noise with respect to fixed timing. (c) Noise on a signal can appear as jitter when compared with a fixed threshold.

the accuracy with which the original analog signal may be represented – a larger number of bits allowing more accurate quantization, resulting in lower noise and distortion. This process is covered further in Chapter 2, and thus will not be discussed further here; suffice it to say that for digital audio word lengths of up to 24 bits may be considered necessary for high sound quality, whereas for digital video a smaller number of bits per sample (typically 8–10) is adequate. Since the number of bits per sample is related directly to the signal-to-noise (S/N) ratio of the system it can be deduced that the S/N ratio required for high quality sound is greater than that required for high quality pictures.

Although the resolution of audio samples may be greater than that of video samples, the sampling rate of a video system is much higher than that required for audio. Thus the total amount of data required per second to represent a moving video picture is considerably greater than that required to represent a sound signal. (In all cases it is assumed that no form of data reduction is used.) This has important implications when considering the requirements for different types of digital interface.

In computer terminology, eight bits is a *byte*, and this is the unit of storage often used in computer systems, even though data may actually be handled in word lengths considerably longer than eight bits. A binary *word* is sometimes confused with a *byte*, but a word can be of virtually any length, whereas a byte may only ever be eight bits. The bit with the greatest 'weight' in a binary word (the leftmost, or the highest power of two, when written down in conventional form) is called the *most significant bit*, and the bit with the least weight ($2^0 = 1$, which is normally the rightmost bit) is called the *least significant bit*. The following example illustrates the point:

Take the eight-bit binary value '01011101':

	MSB							LSB
Binary	0	1	0	1	1	1	0	1
Binary weight	2^7	2^6	2^5	2^4	2^3	2^2	2^1	2^0
Decimal weight	128	64	32	16	8	4	2	1
Decimal equivalent	0 +	64 +	0 +	16 +	8 +	4 +	0 +	1 = 93

A kilobit (Kb) is 1024 bits (2^{10} bits), and a kilobyte (Kbyte) is 1024 bytes. A megabit (Mb) is 1024 kilobits. Confusingly, in communications terminology a data rate of a kilobit per second represents 1000 bits per second, not 1024 bits per second.

1.4 Serial and parallel communications

The bits of a binary word may be transmitted either in parallel or serial form (see Figure 1.5). In the parallel form each bit is carried over a separate communications channel and the result is at least as many channels as there are bits in the word (there are normally additional lines for controlling the exchange of data on a parallel interface). Thus a 24-bit parallel interface would require at least 24 wires, an earth return, a clock line, and a number of address and handshaking lines. Such an approach is normally used for short distance communications ('buses') within a digital device, but is bulky and uneconomical for use over longer distances.

When data is carried serially it only requires a single channel (although electrically that channel may consist of more than one wire), and this makes it economical and simple to implement over large distances. On a serial interface the bits of a word are sent one after the other, and thus it tends to be slower than the parallel equivalent, but the two are so different that to say this is really oversimplifying the matter since there are some extremely fast serial interfaces around. Some serial interfaces carry clock and control information over the same channel as the data, whereas others accompany the data channel with a number of additional parallel lines and a clock signal to control the flow of data between devices. Depending on the standard protocol in use it is possible for serial data to be sent either MSB first or LSB first (see section 1.3), and knowing the convention is clearly important when interpreting received data. These matters are examined further in the next section.

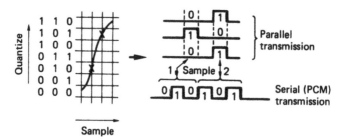

Figure 1.5 When a signal is carried in numerical form, either parallel or serial, the mechanisms of Figure 1.4 ensure that the only degradation is in the conversion process.

1.5 Introduction to interface terminology

It is important to understand certain fundamental principles which will arise regularly in the discussion of interfacing and communications.

1.5.1 Data rate versus baud rate

The data rate of an interface is the rate at which information is carried (sometimes referred to as the information rate), whereas the baud rate of an interface is the *modulation* rate or number of data 'symbols' per second. Although in many cases the two are equivalent, modulation schemes exist which allow for more than one bit to be carried per baud, such as by using multi-level encoding, by using a form of phase-shift keying or other such channel code (see section 3.6).

Data rate is normally quoted as so many kilo- or megabits per second (Kb/s, Mb/s) and this must normally include any capacity for control and additional data. The term *baud rate* is used more widely in computer and telecommunications systems than it is in audio and video interfacing, but it is useful to understand the distinction.

1.5.2 Synchronous, asynchronous and isochronous communications

The receiving device must be able to determine in which time slot it is to register each bit of data which arrives and there are two approaches to achieving this end. In synchronous communications a clock signal normally accompanies the data, either on a separate wire or modulated with the data (see Chapter 3) and this is used to synchronize the receiver's clock to that of the transmitter. Each bit of data may be latched at the receiver on one of the edges of a separate clock, or the clock may be extracted from the modulated data using a suitable phase-locked loop (PLL), as described in section 3.5.

In asynchronous communications the clocks of the transmitter and receiver are not locked directly, but must have an almost identical frequency. The tolerance is often around ±1%. In such a protocol each byte of data is prefixed with a start bit and followed by one or more stop bits (see Figure 1.6) and the phase of the receiver's clock is adjusted at the trailing edge of the start bit. The following data bits are then clocked in according to the receiver's clock, which should remain sufficiently in phase with the transmitted data over the duration of one byte to ensure correct reception. The receiver's clock is then resynchronized at the start of the next data byte. Such an approach is often used in computer systems for exchange of data with remote locations over a modem, for example, where the gaps between received bytes may be variable and data flow may not be regular.

In isochronous systems one master clock is used to synchronize all receiving devices, and data transmission and reception is clocked with relation to this source.

1.5.3 Uni- and bi-directional interfaces

In a uni-directional interface data may only be transmitted in one direction, and no return path is allowed from the receiver back to the transmitter. In a bi-directional interface a return path is provided and this allows for two-way communications. The return path is often used in a simple serial situation to send back handshaking information to the transmitter, telling it whether or not the data was received satisfactorily. Unfortunately such an approach is not particularly useful in digital audio and video interfacing

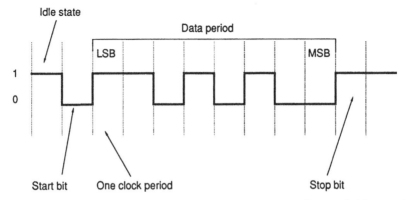

Figure 1.6 In asynchronous data transfer each data byte is normally preceded by a start bit and followed by a stop bit to synchronize the receiver.

because the data in these situations is transferred in real time, making retransmission in the case of an error an unrealistic concept.

A simplex interface is one which operates in one direction only; a half-duplex interface is one which operates in both directions, but only one at a time; and a full-duplex interface is one capable of simultaneous transmission and reception.

1.5.4   Clock signals

A clock signal may be needed, as described above, to synchronize the receiver. In digital audio and video interfacing a variety of methods are used to ensure synchronism between transmitter and receiver which are discussed in detail in the sections concerned. In general two important clock frequencies exist, that is the 'word clock' which indicates the sampling frequency or rate of sample words over the interface, and the 'bit clock' which indicates the rate of individual data bits. Some interfaces, such as the AES/EBU audio interface, combine the bit clock with the data using a modulation scheme or channel code known as bi-phase mark and indicate the starts of sample words using a violation of the modulation scheme which is easily detected by the receiver. This approach avoids the need for additional lines to carry clock signals and the data is said to be 'self-clocking'. In contrast, an interface such as Mitsubishi's audio interface carries the bit clock and word clock signals on individual wires.

An alternative approach is that used in the so-called 'MADI' audio interface (see Chapter 4) in which the transmitter and receiver are both locked to a common reference clock signal and the data is transmitted asynchronously, using a buffer at both ends of the interface to allow for flexibility in timing. This is a form of isochronous approach. The different methods are summarized in Figure 1.7.

1.5.5   Multiplexing

A multiplexed interface is one which carries more than one data signal over a single channel. This is normally achieved using time-division multiplexing (TDM), whereby different time slots in the data stream carry different kinds of information (see Figure 1.8). Here the channel capacity is divided up between the data streams using it, and a multiplexer takes control of the insertion of data into the correct time slots. A demultiplexer at the receiving end extracts the data from each time slot and produces a number of separate data streams.

The AES/EBU audio interface is a simple example of a multiplexed interface, since left and right channel data are multiplexed over the same

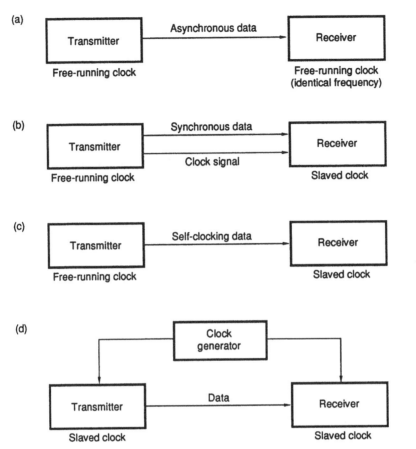

Figure 1.7 A number of approaches may be used to ensure synchronization in data transfer. (a) Asynchronous transfer relies on transmitter and receiver having identical frequencies, requiring only the receiver's clock phase to be adjusted by the start bit of each byte. (b) In synchronous transfer the data signal is accompanied by a separate clock. (c) A form of synchronous transfer involves modulating the data in such a way that a clock signal is part of the channel code. (d) In the isochronous approach both devices are locked to a common clock.

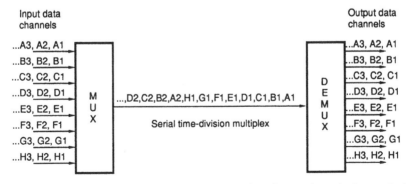

Figure 1.8 In a serial time-division multiplex a number of input channels share a single high-speed link. Each time slot carries a data packet for each channel.

communications channel, with samples of data for each channel taking alternate time slots on the interface. The MADI interface is a more complicated example in which 56 channels of audio data are multiplexed onto one communications channel – a result of which is that the data rate of the communications channel must be extremely high in order to ensure that data for each audio channel can still be carried in real time.

Thus a multiplexed interface must have a data rate high enough to handle the total data rate of all the data streams which are to be multiplexed onto it, otherwise delays will arise as data queues to use the link. In a computer network it is often acceptable to have short delays where the network is shared between many users, since one can wait for a file to load from a remote server, but for real-time applications such as audio and video it is clearly unacceptable to have a shared network which cannot always carry each channel without breaks or queuing.

1.5.6 Buffering

Buffering is sometimes used at transmitting and receiving ends of an interface to store a number of samples of data temporarily. The buffer is normally a RAM (Random Access Memory) store configured in the FIFO (First In, First Out) mode whose input may be addressed separately to its output (see Figure 1.9). Using such a buffer it is possible to iron out irregularities in data flow, since erratic data arriving at the input may be stored in the buffer and read out at a more regular rate after a short delay. The approach will be successful provided that the average rate of flow into the buffer equals the average rate of flow out of it, and the buffer is big enough to accommodate the irregularities which may arise at its input, otherwise the buffer will either overflow or become empty after a time.

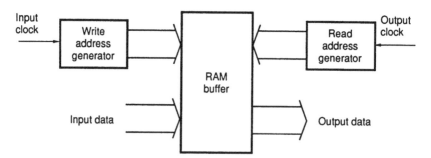

Figure 1.9   A memory buffer may be used as a temporary store to handle irregularities in data flow. Data samples are written to successive memory locations and read out a short time later under control of the read clock.

A disadvantage of the approach is the delay which arises between input and output, which may be undesirable.

1.6 Introduction to networks

In the most general sense a network is a means of communication between a large number of places. According to this definition the Post Office is a network, as are parcel and courier companies. This type of network delivers physical objects. If, however, we restrict the delivery to information only the result is a telecommunications network. The telephone system is a good example of a telecommunications network because it displays most of the characteristics of later networks.

It is fundamental in a network that any port can communicate with any other port. Figure 1.10 shows a primitive three-port network. Clearly each port must select one or other of the remaining ports in a trivial switching system. However, if it were attempted to redraw Figure 1.10 with 100 ports, each one would need a 99-way switch and the number of wires needed would be phenomenal. Another approach is needed.

Figure 1.11 shows that the common solution is to have an exchange, also known as a router, hub or switch, which is connected to every port by

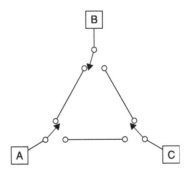

Figure 1.10 A simple three-port network has trivial switching requirements.

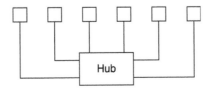

Figure 1.11 A network implemented with a router or hub.

a single cable. In this case when a port wishes to communicate with another, it instructs the switch to make the connection. The complexity of the switch varies with its performance. The minimal case may be to install a single input selector and a single output selector. This allows any port to communicate with any other, but only one at a time. If more simultaneous communications are needed, further switching is needed. The extreme case is where every possible pair of ports can communicate simultaneously.

The amount of switching logic needed to implement the extreme case is phenomenal and in practice it is unlikely to be needed. One fundamental property of networks is that they are seldom implemented with the extreme case supported. There will be an economic decision made balancing the number of simultaneous communications with the equipment cost. Most of the time the user will be unaware that this limit exists, until there is a statistically abnormal condition which causes more than the usual number of nodes to attempt communication.

The phrase 'the switchboard was jammed' has passed into the language and stayed there despite the fact that manual switchboards are only seen in museums. This is a characteristic of networks. They generally only work up to a certain throughput and then there are problems. This doesn't mean that networks aren't useful, far from it. What it means is that with care, networks can be very useful, but without care they can be a nightmare.

There are two key factors to get right in a network. The first is that it must have enough throughput, bandwidth or connectivity to handle the anticipated usage and the second is that a priority system or algorithm is chosen which has appropriate behaviour during overload. These two characteristics are quite different, but often come as a pair in a network corresponding to a particular standard.

Where each device is individually cabled, the result is a radial network shown in Figure 1.12(a). It is not necessary to have one cable per device and several devices can co-exist on a single cable if some form of multiplexing is used. This might be time-division multiplexing (TDM) or frequency division multiplexing (FDM). In TDM, shown in Figure 1.12(b), the time axis is divided into steps which may or may not be equal in length. In Ethernet, for example, these are called frames. During each time step or frame a pair of nodes have exclusive use of the cable. At the end of the time step another pair of nodes can communicate. Rapidly switching between steps gives the illusion of simultaneous transfer between several pairs of nodes. In FDM, simultaneous transfer is possible because each message occupies a different band of frequencies in the cable. Each node has to 'tune' to the correct signal. In practice it is possible to combine FDM and TDM. Each frequency band can be time multiplexed in some applications.

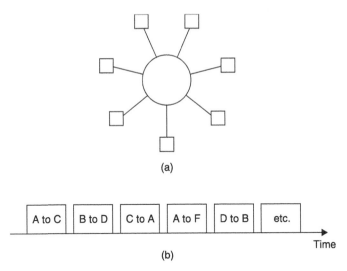

(a)

(b)

Figure 1.12 Radial network at (a) has one cable per node. TDM network (b) shares time slots on a single cable.

Data networks originated to serve the requirements of computers and it is a simple fact that most computer processes don't need to be performed in real time or indeed at a particular time at all. Networks tend to reflect that background as many of them, particularly the older ones, are asynchronous.

Asynchronous means that the time taken to deliver a given quantity of data is unknown. A TDM system may chop the data into several different transfers and each transfer may experience delay according to what other transfers the system is engaged in. Ethernet and most storage system buses are asynchronous. For broadcasting purposes an asynchronous delivery system is no use at all, but for copying a video data file between two storage devices an asynchronous system is perfectly adequate.

The opposite extreme is the synchronous system in which the network can guarantee a constant delivery rate and a fixed and minor delay. An AES/EBU router is a synchronous network.

In between asynchronous and synchronous networks reside the isochronous approaches. These can be thought of as sloppy synchronous networks or more rigidly controlled asynchronous networks. Both descriptions are valid. In the isochronous network there will be maximum delivery time which is not normally exceeded. The data transmission rate may vary, but if the rate has been low for any reason, it will accelerate to prevent the maximum delay being reached. Isochronous networks can deliver near-real-time performance. If a data buffer is provided at both ends, synchronous data such as AES/EBU audio can be fed through an isochronous network. The magnitude of the maximum delay determines the size of the buffer and the length of the fixed overall delay through the system. This delay is responsible for the term 'near-real time'. ATM is an isochronous network.

These three different approaches are needed for economic reasons. Asynchronous systems are very efficient because as soon as one transfer completes, another can begin. This can only be achieved by making every device wait with its data in a buffer so that transfer can start immediately. Asynchronous systems also make it possible for low bit rate devices to share a network with high bit rate devices. The low bit rate device will only need a small buffer and will therefore send short data blocks, whereas the high bit rate device will send long blocks. Asynchronous systems have no difficulty in handling blocks of varying size, whereas in a synchronous system this is very difficult.

Isochronous systems try to give the best of both worlds, generally by sacrificing some flexibility in block size. FireWire (see Chapter 5) is an example of a network which is part isochronous and part asynchronous so that the advantages of both are available.

Whilst computer industry network technology is not ideal for real-time audio-visual information, it has to be accepted that the volume of computer equipment production is such that this technology must be less expensive than hardware specifically designed for broadcast use. There will thus be economic pressure to find ways of using it for audio-visual applications.

1.7 The electrical interface

Although digital interfaces are primarily concerned with the transfer of binary data, there are 'analog' problems to be considered, since the electrical characteristics of the interface such as the type of cable used, its frequency response and impedance will affect the ability of the interface to carry data signals over distances without distortion.

1.7.1 Balanced and unbalanced compared

In an unbalanced interface there is one signal wire and a ground, and the data signal alternates between a positive and negative voltage with respect to ground (see Figure 1.13). The shield of the cable is normally connected to the ground at the transmitter end, and may or may not be connected at the receiver end depending on whether there is a problem with earth loops (a situation in which the earths of the two devices are at different potentials, causing a current to circulate between them, sometimes resulting in hum induction into the signal wire). The unbalanced interface tends to be quite susceptible to interference, since any unwanted signal induced in the data wire will be inseparable from the wanted signal.

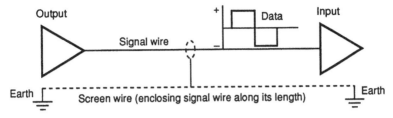

Figure 1.13 Electrical configuration of an unbalanced interface.

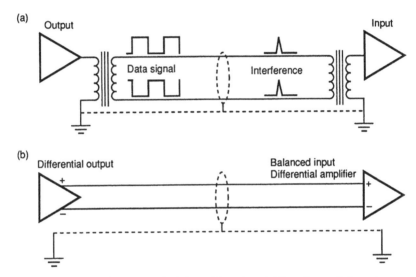

Figure 1.14 Electrical configuration of a balanced interface. (a) Transformer balanced. (b) Electronically balanced.

In a balanced interface there are two signal wires and a ground (see Figure 1.14) and the interface is terminated at both ends either in a differential amplifier or a transformer. The driver drives the two legs of the line in opposite phase, and the advantage of the balanced interface is that any interfering signal is induced equally into the two legs, *in phase*. At the receiver any so-called 'common mode' signals are cancelled out either in the transformer or differential amplifier, since such devices are only interested in the *difference* between the two legs. The degree to which the receiver can reject common mode signals is called the common mode rejection ratio (CMRR). Although it is a generally held belief that transformers offer better CMR than electronically balanced lines, the performance of modern differential line drivers and receivers is often as good. The advantage of transformers is that they make the line truly 'floating', that is independent of the ground, and there is no DC coupling across them, thus isolating the two devices.

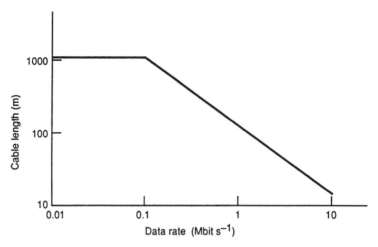

Figure 1.15 The RS-422 standard allows different cable lengths at different frequencies. Shown here is the guideline for data signalling rate versus cable length when using twisted-pair cable with a wire diameter of 0.51 mm. Longer distances may be achieved using thinner wire.

The balanced interface therefore requires one more wire than the unbalanced interface, and will usually have lines labelled 'Ground', 'Data+' and 'Data−', whereas the unbalanced interface will simply have 'Ground' and 'Data'. For temporary test set-ups it is sometimes possible to interconnect between balanced and unbalanced electrical interfaces or vice versa, by connecting the unbalanced interface between the two legs of the balanced one, or between the ground and one leg, but often the voltages involved are different, and one must take care to ensure that the two data streams are compatible. Balancing transformers are available which will convert a signal from one form to the other.

1.7.2 Electrical interface standards

The different interface standards specify various peak-to-peak voltages for the data signal, and also specify a minimum acceptable voltage at the receiver to ensure correct decoding (this is necessary because the signal may have been attenuated after passing over a length of cable). Quite commonly serial interfaces conform to one of the international standard conventions which describe the electro-mechanical characteristics of data interfaces, such as RS-422[5] which is a standard for balanced communication over long lines devised by the EIA (Electronics Industries Association). For example, the AES3-1992 audio interface is designed to be able to use RS-422 drivers and receivers and specifies a peak-to-peak voltage between 2 and 7 volts at the transmitter (when measured across a 110 ohm resistor with no interconnecting cable present), and also specifies a minimum 'eye-height' (see section 3.5) at the receiver of 200 mV.

(In passing it should be noted that standards such as RS-422 are mostly only electrical or electro-mechanical standards, and do not say anything about the format or protocol of the data to be carried over them.)

Unbalanced serial interfaces such as RS-232 are not used in audio and video systems, since they are not designed for the high data rates and long distances involved, and are more suited to telecommunications. The voltages involved in an RS-232 interface can be up to 25 V, and thus it is not recommended that one should interconnect an RS-232 output with an RS-422 input (which may be damaged by anything above around 12 volts). An RS-422 interface is designed to carry data at rates up to 100 kbaud over a distance of 1200 metres. Above this rate the distance which may be covered satisfactorily drops with increasing baud rate, as shown in Figure 1.15, depending on whether the line is terminated or not (see section 1.7.3).

Other manufacturer-specific interfaces often use TTL levels of 0–5 volts over unbalanced lines, and these are only suitable for communications over relatively short distances.

1.7.3 Transmission lines

At low data rates a piece of wire can be considered as a simple entity which conducts current and perhaps attenuates the signal resistively to some extent, and in which all components of the signal travel at the same speed as each other, but at higher rates it is necessary to consider the interconnect as a 'transmission line' in which reflections may be set up and where such factors as the characteristic impedance and terminating impedance of the line become important.

When considering a simple electrical circuit it is normal to assume that changes in voltage and current occur at the same time throughout the circuit, since the speed at which electricity travels down a wire is fast enough for this to be a reasonable assumption in many cases. When very long cables are involved, the time taken for an electrical signal to travel from one end to the other may approach a significant proportion of one cycle of the signal's waveform, and when the frequency of the electrical signal is high this is yet more likely since the cycle is short. Another way of considering this is to think in terms of the effective wavelength of the signal in its electrical form, which will be very long since the speed at which electricity travels in wire approaches the speed of light. When the wavelength of the signal in the wire becomes of the same order of magnitude as the length of the wire, transmission line issues may arise.

In a transmission line the ends of the line may be considered to be impedance discontinuities, that is points at which the impedance of the line changes from the line's *characteristic impedance* to the impedance of the *termination*. The characteristic impedance of a line is a difficult concept to grasp but it may be modelled as shown in Figure 1.16, being a combination of inductance and capacitance (and probably, in reality, also some resistance) which depends on the spacing between the conductors, their size and the type of insulation used. Characteristic impedance has been defined as the input impedance of a line of infinite length[6]. The situation at the ends of a transmission line may be likened to what happens when a sound wave hits a wall – a portion of the power is reflected and a portion is absorbed. The wall represents an impedance discontinuity, and at such points sound energy is reflected. At certain frequencies standing wave modes will be set up in the room, at frequencies where multiples of half a wavelength equal one of the dimensions of the room, whereby the reflected wave combines constructively with the incident wave to produce points of maximum and minimum sound pressure within the room.

If an electrical transmission line is incorrectly terminated, reflections will be set up at the ends of the line, resulting in a secondary electrical wave travelling back down the line. This reflected wave may interfere with the transmitted wave, and in the case of a data signal may corrupt it if the reflected energy is high. The mismatch also results in a loss of power transferred to the next stage. If the line is correctly terminated the end of the line will not appear to be a discontinuity, the optimum power transfer will result at this point, and no reflections will be set up. A line is correctly terminated by ensuring that source and receiver impedances are the same as the characteristic impedance of the line (see Figure 1.17).

The upshot of all this for the purposes of this book is that long interconnects which carry high frequency audio or video data signals, often having bandwidths of many megahertz, may be subject to transmission line phenomena, and thus lines should normally be correctly terminated. The penalty for not doing so may be reduced signal strengths and erroneous

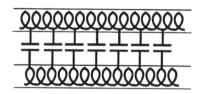

Figure 1.16 Electrical model of a transmission line.

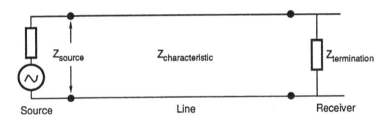

Matched transmission line when $Z_{source} = Z_{characteristic} = Z_{termination}$

Figure 1.17 Electrical characteristics of a matched transmission line.

data reception after some distance, and such situations can sometimes arise, especially when audio data signals are sent down existing cable runs which may pass through jackfields and different types of wire, each of which presents a change in characteristic impedance. In video environments people tend to be used to the concepts of transmission lines and correct termination, since video signals have always been subject to transmission line phenomena. In audio environments it may be a new concept, since analog audio signals do not contain high enough frequencies for such matters to become a problem, yet digital audio signals may. It is not recommended, for example, to parallel a number of receivers across one source line when dealing with digital signals, since the line will not then be correctly terminated and the signal level may also be considerably attenuated.

1.7.4 Cables

The types of cables used in serial audio and video interfaces vary from balanced screened, twisted pair (e.g. AES/EBU) to unbalanced coaxial links (e.g. Sony SDIF-2) and the characteristic impedances of these cables are different. Typical twisted pair audio cable, such as is used in many analog installations (and also put to use in digital links), tends to have a characteristic impedance of around 90–100 ohms, although this is not carefully controlled in manufacture since it normally does not matter for analog audio signals, whilst the more expensive 'star-quad' audio cable has a lower characteristic impedance of around 35 ohms. Typical coaxial cables used in video work have a characteristic impedance of 75 ohms, and other RF coaxial links use 50 ohm cable.

One problem with the twisted pair balanced line is that its electromagnetic radiation is poor compared with that of the coaxial link. A further advantage of the coaxial link is that its attenuation does not become severe until much higher frequencies than that of the twisted pair, and the speed of propagation along a coaxial link is significantly faster than along a twisted pair, resulting in smaller signal delays. But the twisted pair link is balanced whereas the coaxial link is not, and this makes it more immune to external interference, which is a great advantage.

Cable losses at high frequencies will clearly reduce the bandwidth of the interconnect, and the effect of this on the data signal is to slow the rise and fall times of the data edges, and to delay the point of transition between one state and the other by an amount which depends on the state of the data in the previous bit cell. The practical result of this is data link timing jitter, as proposed by Dunn[7], which may affect signal quality in D/A conversion if the clock is recovered from a portion of the data frame which is subject to a large degree of jitter. Links which suffer HF loss can often be equalized at the receiver to prop up the high frequency end of the spectrum, and this can help to accommodate longer interconnects which would otherwise fail. This matter is examined further in section 6.4.1.

Thus a number of factors combine to suggest that the type of cabling used in a digital interconnect is an important matter. A cable is required whose characteristic impedance matches that of the driver and receiver as closely as possible and is consistent along the length of the cable. The cable should have low loss at high frequencies, and the precise specification requires a knowledge of the bandwidth of the signal to be transferred. Cable manufacturers can usually quote such figures in specification sheets or on request, and it pays to study such matters, especially when recabling a large installation. More detailed discussion of these matters is contained within the sections of this book covering individual interfaces.

1.7.5 Connectors

The connectors used in digital interfacing fall into a number of distinct categories (see Figure 1.18). First, there are the unbalanced coaxial connectors, normally BNC type, which differ slightly depending on the characteristic impedance of the line; 50 ohm BNC connectors have a slightly larger central pin than 75 ohm connectors and can damage 75 ohm sockets if used inadvertently. RCA phono connectors, such as are often found in consumer hi-fi systems, are also used for unbalanced consumer digital audio interfaces using coaxial cable, although they are not proper coaxial connectors and do not have a controlled characteristic impedance. Both these connector types carry the data signal on the central pin and the shield on the outer ring.

The XLR-3 connector is used for one balanced digital audio interface, and it has its roots as an analog professional audio connector. The convention is for pin 1 to be the shield, pin 2 to be 'Data+', and pin 3 to be 'Data−'.

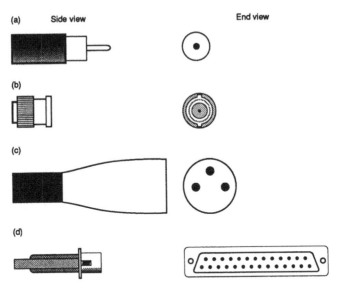

Figure 1.18 A number of connector types are commonly used in digital interfacing. (a) RCA phono connector. (b) BNC connector. (c) XLR connector. (d) D-type connector.

The D-sub type of connector stems from computer systems and remote control applications, and it has a number of individual pins arranged in two or three rows. The 9 pin D connector is often used for RS-422 communications, since it allows for two balanced sends and returns plus a ground, whereas some custom digital interfaces (either parallel or multi-channel serial) use 25, 36 or even 50 pin D-type connectors.

These are not the only connectors used in audio and video interfacing, but they are the most common. Miscellaneous manufacturer-specific formats use none of these, Yamaha preferring the 8 pin DIN connector, for example, as its digital 'cascade' connector in one format.

1.8 Optical interfaces

Optical fibres are now playing a larger part in everyday data communications, and the cost : performance ratio of such fibre interconnects makes them a reasonable proposition when a large amount of data is to be carried over long distances. The key features of optical fibres are a large bandwidth, very low losses and immunity to interference, coupled with small size and flexibility. Also, when fibres are used as interconnects, devices are electrically isolated thus avoiding problems such as ground loops, shorts and crosstalk. Data is transferred over fibres by modulating a light source with the data, the light being carried within the fibre to an optical detector at the receiving end. The light source may be an LED or a laser diode, with the LED capable of operation up to a maximum of a few hundred megahertz at low power, whilst the laser is preferable in applications at higher frequencies or over longer distances.

1.8.1   Fibre principles

The typical construction of an optical fibre is shown in Figure 1.19, and light travels in the fibre as in a typical 'waveguide', by reflection at the boundaries. Total internal reflection occurs at the boundaries between the fibre core and its cladding, provided that the angle at which the light wave hits the boundary is shallower than a certain value called the *critical angle*, and thus the method of coupling the light source to the fibre is important in ensuring that light is propagated in the right way. Unless the fibre is exceptionally narrow, with a diameter approaching the wavelength of the light travelling along it (say between 4 and 10 μm), the light will travel along a number of paths – known as 'multimodes' – and thus the time taken for a source pulse to arrive at the receiver may vary slightly depending on the length of each path. This results in smearing of pulses in the time domain, and is known as *modal dispersion*, or, looked at in the frequency domain, it represents a reduction in the bandwidth of the link with increasing length. Up to distances of around 1.5 km the bandwidth decreases roughly linearly with distance (quoted in MHz per km), and after this in proportion to the square root of the length. Optical signals will also be attenuated with distance due to scattering by metal ions within the fibre, and by absorption due to water present within the structure. Losses are usually quoted in dB per km at a specified wavelength of light, and can be as low as 1 dB/km with high quality silica, graded index multimode fibres, or higher than 100 dB/km with plastic or ordinary glass cores. Clearly the high loss cables would be cheaper, and perhaps adequate for consumer applications in which the distances to be covered might be quite small.

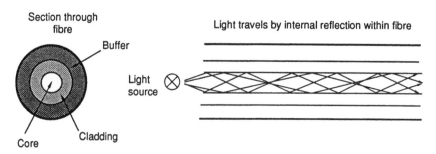

Figure 1.19   Cross-section through a typical optical fibre, and mode of transmission.

Single mode fibres with very fine cores achieve very wide bandwidths with very low losses, and thus are suitable for use over long distances. Attenuations of around 0.5 dB/km are not uncommon with such fibres, which have only recently become feasible due to the development of suitable sources and connectors.

1.8.2 Light sources and connectors

LED (Light-Emitting Diode) light sources are made of gallium arsenide (GaAs) and can be doped to produce light with a wavelength between 800 and 1300 nm. The bandwidth of the radiated light is fairly wide, having a range of wavelengths of around 40 nm, which is another factor leading to greater losses over distance as the light of different wavelengths propagates over different modal paths within the fibre. The light from an LED source is incoherent (i.e. the phase and plane of the wavelets is random), and the angle over which it is radiated is quite wide. Since light radiated into the fibre at angles greater than a certain 'acceptance angle' will not be internally reflected it will effectively be lost in the cladding; thus the effectiveness of the coupling of light from an LED into the fibre is not good and only a few hundred microwatts of power can be transmitted.

An ILD (Injection Laser Diode) on the other hand produces coherent light of a similar wavelength to the LED, but over a narrower angle and with a narrower bandwidth (between around 1 and 3 nm in wavelength), thus providing better coupling of the light power to the fibre and resulting in less dispersion. Because of this, ILD drivers can be used for links which work in the gigahertz region whilst maintaining low losses.

Optical detectors are forms of photodiode which are very efficient at converting received light power into electrical current. Rise times of around 10 ns or less are achievable. The main problem with detectors is the distortion and noise they introduce, and different types of photodiode differ enormously in their S/N ratios. The so-called *avalanche photodiode* has a noise floor considerably lower than its counterpart, the *PIN diode*, and thus is preferable in critical applications. The *integrated detector/preamplifier* (IPD) provides amplification and detection of the light source in an integrated device, providing a higher output level than the other two and an improved S/N ratio.

The important features of connectors are their insertion loss and their return loss, these being respectively the amount of power lost due to the insertion of a connector into an otherwise unbroken link, and the amount of power reflected back in the direction of the source due to the presence of the connector. Typically insertion loss should be low (less than 1 dB), and return loss should be high (greater than 40 dB), for a reliable installation.

There are seven principal types of fibre optic connector, and it is not intended to cover each of them in detail here. For a comprehensive survey the reader is referred to Ajemian[8]. The so-called FDDI (Fibre Distributed Digital Interface) MIC connector is gaining widespread use in optical networks since it is a duplex connector (allows separate communication in two directions) with a typical insertion loss of 0.6 dB, whilst the SC connector is a popular 'snap-on' device developed by the Japanese NTT Corporation, available in both simplex and duplex forms, offering low insertion loss of around 0.25 dB and small size. The ST series of connectors, developed by AT&T, is also used widely.

1.9   Timebase recovery in interfacing

Unlike generic data, audio-visual data is only meaningful when reproduced with an appropriate timebase, both in the long term and the short term. Whatever means may be used to deliver the data, there must also be a means to recover the timebase. In audio and video installations it is common practice to distribute master clocks from a central generator. When this is done, the interfaces essentially become data transmissions, because timebase correction will be used at any receiving device in order to align the signal timing with the master reference.

This approach is feasible within a building complex, but not over long distances. Where data networks are used, packets may be routed over different physical paths between the same points and the time taken will be subject to some variation, known as packet jitter. Under these conditions it is much more difficult to recreate the correct timebase in the receiving device. Where several signal formats are possible, the receiver may need metadata that help it generate the correct timing.

References

1. Rumsey, F.J., *Desktop Audio Technology*, Focal Press, Oxford (2004)
2. Watkinson, J.R., *The Art of Digital Audio*, third edition, Focal Press, Oxford (2001)
3. Watkinson, J.R., *The Art of Digital Video*, third edition, Focal Press, Oxford (2000)
4. Connor, F.R., *Modulation*, Edward Arnold, London (1982)
5. EIA., *Industrial electronics bulletin no. 12*. EIA standard RS-422A. Electronics Industries Association, Engineering Dept., Washington DC
6. Sinnema, W., *Electronic Transmission Technology*, Prentice-Hall, Englewood Cliffs, NJ (1979)
7. Dunn, J., Jitter: specification and assessment in digital audio equipment. Presented at the *93rd AES Convention, San Francisco*, 1–4 October. Audio Engineering Society (1992)
8. Ajemian, R.G., Fiber-optic connector considerations for professional audio. *Journal of the Audio Engineering Society*, vol. 40, no. 6, June, pp. 524–531 (1992)

2

An introduction to digital audio and video

2.1 What is an audio signal?

Actual sounds are converted to electrical signals for convenience of handling, recording and conveying from one place to another. This is the job of the microphone. There are two basic types of microphone: those which measure the variations in air pressure due to sound, and those which measure the air velocity due to sound, although there are numerous practical types which are a combination of both.

The sound pressure or velocity varies with time and so does the output voltage of the microphone, in proportion. The output voltage of the microphone is thus an analog of the sound pressure or velocity.

As sound causes no overall air movement, the average velocity of all sounds is zero, which corresponds to silence. As a result the bi-directional air movement gives rise to bipolar signals from the microphone, where silence is in the centre of the voltage range, and instantaneously negative or positive voltages are possible. Clearly the average voltage of all audio signals is also zero, and so when level is measured, it is necessary to take the modulus of the voltage, which is the job of the rectifier in the level meter. When this is done, the greater the amplitude of the audio signal, the greater the modulus, and so a higher level is displayed.

2.2 Types of audio signal

Whilst the nature of an audio signal is very simple, there are many applications of audio, each requiring different bandwidth and dynamic range.

Like video signals, audio signals exist in a variety of formats. Audio can mean a single monophonic signal, or several signals can be required to deliver sound with spatial attributes. In general timing and level errors between these signals will be detrimental. In stereo, two channels are required to give a spread of virtual sound sources between a pair of loud-speakers. With more channels, loudspeakers can be provided at the rear and sides to give a feeling of ambience.

A common surround sound format has five channels, where one of these feeds a front centre loudspeaker. This centre channel is not strictly necessary but exists because of the rather different circumstances in which cinema sound has traditionally been produced. In some cases a sixth channel is used for low frequency effects.

In the digital domain it is straightforward to deliver several discrete audio channels, whereas in analog systems it is not. As a result systems such as Dolby Surround were developed to allow some surround effect to be encoded into only two analog channels.

2.3   What is a video signal?

The goal of television is to allow a moving picture to be seen at a remote place. The picture is a two-dimensional image, which changes as a function of time. This is a three-dimensional information source where the dimensions are distance across the screen, distance down the screen and progression in time.

Whilst telescopes convey these three dimensions directly, this cannot be done with electrical signals or radio transmissions, which are restricted to a single parameter varying with time.

The solution in film and television is to convert the three-dimensional moving image into a series of still pictures, taken at the frame rate, and then, in television only, the two-dimensional images are scanned as a series of lines to produce a single voltage varying with time which can be recorded or transmitted.

2.4   Types of video signal

Figure 2.1 shows some of the basic types of analog colour video; each of these types can, of course, exist in a variety of line standards. Since practical colour cameras generally have three separate sensors, one for each primary colour, an RGB system will exist at some stage in the internal workings of the camera, even if it does not emerge in that form.

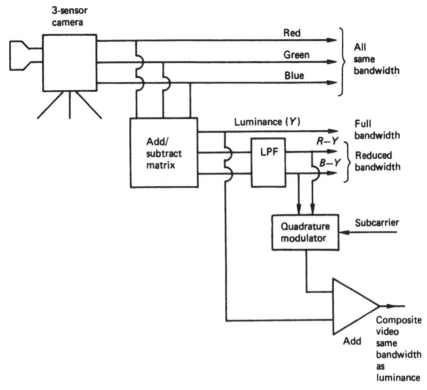

Figure 2.1 The major types of analog video. Red, green and blue signals emerge from the camera sensors, needing full bandwidth. If a luminance signal is obtained by a weighted sum of R, G and B, it will need full bandwidth, but the colour difference signals $R-Y$ and $B-Y$ need less bandwidth. Combining $R-Y$ and $B-Y$ in a subcarrier modulation scheme allows colour transmission in the same bandwidth as monochrome.

RGB consists of three parallel signals each having the same spectrum, and is used where the highest accuracy is needed, often for production of still pictures. Examples of this are paint systems and in computer-aided design (CAD) displays. RGB is seldom used for real-time video recording; there is no standard RGB recording format for post-production or broadcast. As the red, green and blue signals directly represent part of the image, this approach is known as component video.

Some saving of bandwidth can be obtained by using colour difference working. The human eye relies on brightness to convey detail, and much less resolution is needed in the colour information. R, G and B are matrixed together to form a luminance (and monochrome compatible) signal Y, which has full bandwidth. The matrix also produces two colour difference signals, $R-Y$ and $B-Y$, but these do not need the same bandwidth as Y; one-half or one-quarter will do depending on the application. In casual parlance, colour difference formats are often called component formats to distinguish them from composite formats.

For colour television broadcast in a single channel, the PAL and NTSC systems interleave into the spectrum of a monochrome signal a subcarrier that carries two colour difference signals of restricted bandwidth. The subcarrier is intended to be invisible on the screen of a monochrome television set. A subcarrier-based colour system is generally referred to as composite video, and the modulated subcarrier is called chroma.

The majority of today's broadcast standards use 2:1 interlace in order to save bandwidth. Figure 2.2(a) shows that in such a system, there are an odd number of lines in a frame, and the frame is split into two fields. The first field begins with a whole line and ends with a half line, and the second field begins with a half line, which allows it to interleave spatially with the first field. Interlace may be viewed as a crude compression technique which was essentially rendered obsolete by the development of digital image compression techniques such as MPEG.

The field rate is intended to determine the flicker frequency, whereas the frame rate determines the bandwidth needed, which is thus halved along with the information rate. Information theory tells us that halving the information rate must reduce quality, and so the saving in bandwidth is accompanied by a variety of effects that are actually compression artefacts. Figure 2.2(b) shows the spatial/temporal sampling points in a 2:1 interlaced

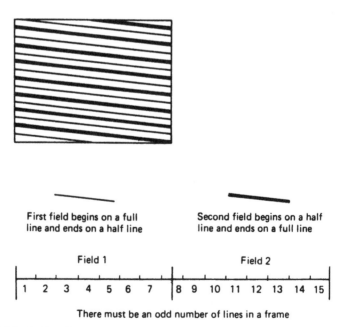

First field begins on a full line and ends on a half line

Second field begins on a half line and ends on a full line

Field 1 | Field 2

1 2 3 4 5 6 7 | 8 9 10 11 12 13 14 15

There must be an odd number of lines in a frame

Figure 2.2(a)   2:1 interlace.

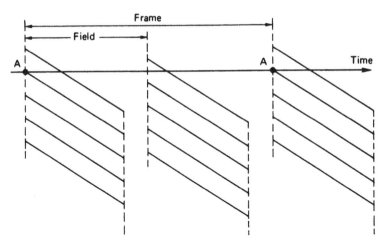

Figure 2.2(b) In an interlaced system, a given point A is only refreshed at frame rate, causing twitter on fine vertical detail.

system. If an object has a sharp horizontal edge, it will be present in one field but not in the next. The refresh rate of the edge will be reduced to frame rate, and becomes visible as twitter. Whilst the vertical resolution of a test card is maintained with interlace, apart from the twitter noted, the ability of an interlaced standard to convey motion is halved. In the light of what is now known[1], interlace causes degradation roughly proportional to bandwidth reduction, and so should not be considered for any future standards.

A wide variety of frame rates exist. Initially it was thought that a frame rate above the critical flicker frequency (CFF) of human vision at about 50 Hz would suffice, but this is only true for still pictures. When a moving object is portrayed, the eye tracks it and thus sees the background presented as a series of stills each in a different place. It is thus the prevention of background strobing that should be the criterion for frame rate. Needless to say the frame rate of film at 24 Hz is totally inadequate and the 50 and 60 Hz rates of television are suboptimal. Equipment running at 75 Hz or more gives obviously more fluid and realistic motion.

Originally film ran at 18 fps and to reduce flicker each frame was shown three times using a multi-blade shutter. During the development of 'talking pictures' it was found that the linear film speed at 18 Hz provided insufficient audio bandwidth and the frame rate was increased to 24 Hz to improve the sound quality. Thus the frame rate of movie film to this day is based on no considerations of human vision whatsoever. The adoption of digital film production techniques that simply retain this inadequate frame rate shows a lack of vision. A digitized film frame is essentially the same as a progressively scanned television frame, but at 24 or 25 Hz, a video monitor will be unacceptable because of flicker. This gave rise to the idea of a 'segmented frame' in which the simultaneously captured frame was displayed as fields containing alternately the odd and even lines. This allows digital film material to be viewed on television production equipment.

European television standards reflect the local AC power frequency and use 50 fields per second. American standards initially used 60 Hz field rate but on the introduction of colour the rate had to be reduced by 0.1% to 59.94 Hz to prevent chroma and sound interference.

Film cameras are flexible devices and where film is being shot for television purposes the film frame rate may be made 25 Hz or even 30 Hz so that two television fields may be created from each film frame in a telecine machine.

With the move towards high definition television, obviously more lines were required in the picture, with more bandwidth to allow more detail along the line. This, of course, only gives a better static resolution, whereas what was needed was better resolution in the case of motion. This requires an increase in frame rate, but none was forthcoming. High definition formats using large line counts but which retained interlace gave poor results because the higher static resolution made it more obvious how poor the resolution became in the presence of motion.

Those high definition standards using progressive scan gave better results, with 720 P easily outperforming interlaced formats. Digital broadcasting systems support the use of progressively scanned pictures. Video interfaces have to accept the shortcomings inherent in the signals they convey. Their job is simply to pass on the signals without any further degradation.

2.5   What is a digital signal?

It is a characteristic of analog systems that degradations cannot be separated from the original signal, so nothing can be done about them. At the end of a system a signal carries the sum of all degradations introduced at each stage through which it passed. This sets a limit to the number of stages through which a signal can be passed before it is useless. Alternatively, if many stages are envisaged, each piece of equipment must be far better than necessary so that the signal is still acceptable at the end. The equipment will naturally be more expensive.

One of the vital concepts to grasp is that digital audio and video are simply alternative means of carrying the same information. An ideal digital recorder has the same characteristics as an ideal analog recorder: both of them are totally transparent and reproduce the original applied waveform without error. One need only compare high quality analog and digital equipment side by side with the same signals to realize how transparent modern equipment can be. Needless to say, in the real world, ideal conditions seldom prevail, so analog and digital equipment both fall short of the ideal. Digital equipment simply falls short of the ideal to a smaller extent than does analog and at lower cost, or, if the designer chooses, can have the same performance as analog at much lower cost.

Although there are a number of ways in which audio and video waveform can be represented digitally, there is one system, known as pulse code modulation (PCM), that is in virtually universal use. Figure 2.3 shows how PCM works. Instead of being continuous, the time axis is represented in a discrete, or stepwise, manner. The waveform is not carried by continuous representation, but by measurement at regular intervals. This process is called sampling and the frequency with which samples are taken is called the sampling rate or sampling frequency F_s. The sampling rate is generally fixed and is not necessarily a function of any frequency in the signal, although in video it may be for convenience. If every effort is

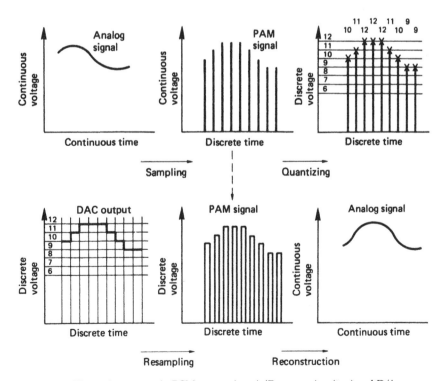

Figure 2.3 The major process in PCM conversion. A/D conversion (top) and D/A conversion (bottom). Note that the quantizing step can be omitted to examine sampling and reconstruction independently of quantizing (dotted arrow).

made to rid the sampling clock of jitter, or time instability, every sample will be made at an exactly even time step. Clearly if there is any subsequent time-base error, the instants at which samples arrive will be changed and the effect can be detected. If samples arrive at some destination with an irregular timebase, the effect can be eliminated by storing the samples temporarily in a memory and reading them out using a stable, locally generated clock. This process is called timebase correction and all properly engineered digital systems must use it. Clearly timebase error is not simply reduced; it can be totally eliminated. As a result there is little point measuring the wow and flutter or timebase error of a digital recorder; it doesn't have any. What happens is that the crystal clock in the timebase corrector measures stability of the measuring equipment. It should be stressed that sampling is an analog process. Each sample still varies infinitely as the original waveform did.

Those who are not familiar with digital processes often worry that sampling takes away something from a signal because it is not taking notice of what happened between the samples. This would be true in a system having infinite bandwidth, but no analog signal can have infinite bandwidth. All analog signal sources from microphones, tape decks, cameras and so on have a frequency response limit, as indeed do our ears and eyes. When a signal has finite bandwidth, the rate at which it can change is limited, and the way in which it changes becomes predictable. When a waveform can only change between samples in one way, it is then only necessary to carry the samples and the original waveform can be reconstructed from them.

Figure 2.3 also shows that each sample is also discrete, or represented in a stepwise manner. The length of the sample, which will be proportional to the voltage of the waveform, is represented by a whole number. This process is known as quantizing and results in an approximation, but the size of the error can be controlled until it is negligible. If, for example, we were to measure the height of humans to the nearest metre, virtually all adults would register 2 metres high and obvious difficulties would result. These are generally overcome by measuring height to the nearest centimetre. Clearly there is no advantage in going further and expressing our height in a whole number of millimetres or even micrometres. The point is that an appropriate resolution can be found just as readily for audio or video, and greater accuracy is not beneficial. The link between quality and sample resolution is explored later in this chapter. The advantage of using whole numbers is that they are not prone to drift. If a whole number can be carried from one place to another without numerical error, it has not changed at all. By describing waveforms numerically, the original information has been expressed in a way that is better able to resist unwanted changes.

Essentially, digital systems carry the original waveform numerically. The number of the sample is an analog of time, and the magnitude of the sample is an analog of the signal voltage. As both axes of the waveform are discrete, the waveform can be accurately restored from numbers as if it were being drawn on graph paper. If we require greater accuracy, we simply choose paper with smaller squares. Clearly more numbers are required and each one could change over a larger range.

In simple terms, the waveform is conveyed in a digital recorder as if the voltage had been measured at regular intervals with a digital meter and the readings had been written down on a roll of paper. The rate at which the measurements were taken and the accuracy of the meter are the only factors which determine the quality, because once a parameter is expressed as a discrete number, a series of such numbers can be conveyed unchanged. Clearly in this example the handwriting used and the grade of paper have no effect on the information. The quality is determined only by the accuracy of conversion and is independent of the quality of the signal path.

In practical systems, binary numbers are used, as was explained in Chapter 1 in which it was also shown that there are two ways in which binary signals can be used to carry samples. When each digit of the binary number is carried on a separate wire this is called parallel transmission. The states of the signals change at the sampling rate. This approach is used in the parallel video interfaces, as video needs a relatively short word length: eight or ten bits. Using multiple wires is cumbersome where a long word length is in use, and a single wire can be used where successive digits from each sample are sent serially. This is the definition of pulse code modulation. Clearly the clock frequency must now be higher than the sampling rate. Whilst the transmission of audio by such a scheme is advantageous in that noise and timebase error have been eliminated, there is a penalty that a single high quality audio channel requires around one million bits per second. Digital audio could only come into use when such a data rate could be handled economically.

As a digital video channel requires of the order of two hundred million bits per second it is not surprising that digital audio equipment became common somewhat before digital video.

2.6 Why digital?

There are two main answers to this question, and it is not possible to say which is the most important, as it will depend on one's standpoint:

(a) The quality of reproduction of a well-engineered digital system is independent of the medium and, in the absence of a compression scheme, depends only on the quality of the conversion processes.

(b) The conversion to the digital domain allows tremendous opportunities that were denied to analog signals.

Someone who is only interested in quality will judge the former the most relevant. If good quality convertors can be obtained, all of the shortcomings of analog recording and transmission can be eliminated to great advantage. One's greatest effort is expended in the design of convertors, whereas those parts of the system that handle data need only be workmanlike. Wow, flutter, timebase error, vector jitter, crosstalk, particulate noise, print-through, dropouts, modulation noise, HF squashing, azimuth error and interchannel phase errors are all history. When a digital recording is copied, the same numbers appear on the copy: it is not a dub, it is a clone. If the copy is indistinguishable from the original, there has been no generation loss. Digital recordings can be copied indefinitely without loss of quality through a digital interface.

In the real world everything has a cost, and one of the greatest strengths of digital technology is low cost. If copying causes no quality loss, recorders do not need to be far better than necessary in order to withstand generation loss. They need only be of adequate quality on the first generation if that quality is then maintained. There is no need for the great size and extravagant tape consumption of professional analog recorders. When the information to be recorded is discrete numbers, they can be packed densely on the medium without quality loss. Should some bits be in error because of noise or dropout, error correction can restore the original value. Digital recordings take up less space than analog recordings for the same or better quality. Tape costs are far less and storage costs are reduced.

Digital circuitry costs less to manufacture. Switching circuitry handling binary can be integrated more densely than analog circuitry. More functionality can be put in the same chip. Analog circuits are built from a host of different component types having a variety of shapes and sizes and are costly to assemble and adjust. Digital circuitry uses standardized component outlines and is easier to assemble on automated equipment. Little if any adjustment is needed. Once audio or video signals are in the digital domain, they become data, and apart from the need to be reproduced at the correct sampling rate, are indistinguishable from any other type of data. Systems and techniques developed in other industries for other purposes can be used for audio and video. Computer equipment is available at low cost because the volume of production is far greater than that of professional audio equipment. Disk drives and memories developed for computers can be put to use in audio and video products. A word processor adapted to handle audio samples becomes a workstation. If video data is handled the result is a non-linear editor. Communications networks developed to handle data can happily carry digital audio and video over indefinite distances without quality loss. Digital audio broadcasting (DAB) makes use of these techniques to eliminate the interference, fading and multipath reception problems of analog broadcasting. At the same time, more efficient use is made of available bandwidth. Digital broadcasting techniques are now being applied to television signals, with DVB being used in Europe and other parts of the world.

Digital equipment can have self-diagnosis programs built in. The machine points out its own failures. Routine, mind-numbing adjustment of analog circuits to counteract drift is no longer needed. The cost of maintenance falls. A small operation may not need maintenance staff at all; a service contract is sufficient. A larger organization will still need maintenance staff, but they will be fewer in number and more highly skilled.

2.7 The information content of an analog signal

Any analog signal source can be characterized by a given useful bandwidth and signal-to-noise ratio. Video signals have very wide bandwidth extending over several MHz and require only 50 dB or so S/N ratio whereas audio signals require only 20 kHz but need a much better S/N ratio. If a well-engineered digital channel having a wider bandwidth and a greater signal-to-noise ratio is put in series with such a source, it is only necessary to set the levels correctly and the signal is then subject to no loss of information whatsoever.

Provided the digital clipping level is above the largest input signal, the digital noise floor is below the inherent noise in the input signal and the low- and high-frequency response of the digital channel extends beyond the frequencies in the input signal, then the digital channel is a 'wider window' than the input signal needs and its extremities cannot be explored by that signal. As a result there is no test known which can reliably tell whether or not the digital system was or was not present, unless, of course, it is deficient in some quantifiable way.

In audio the wider-window effect is obvious on certain Compact Discs made from analog master tapes. The CD player faithfully reproduces the tape hiss, dropouts and HF squashing of the analog master, which render the entire CD mastering and reproduction system transparent by comparison.

On the other hand, if an analog source can be found which has a wider window than the digital system, then the presence of the digital system will be evident either due to the reduction in bandwidth or the reduction in dynamic range. This will be evident on listening to DAB broadcasts that use compression and are inferior to the best analog practice.

2.8 Introduction to conversion

There are a number of ways in which an audio waveform can be digitally represented, but the most useful and therefore common is PCM, which was introduced in Chapter 1. The input is a continuous-time, continuous-voltage waveform, and this is converted into a discrete-time, discrete-voltage format by a combination of sampling and quantizing. As these two processes are orthogonal (at right angles to one another) they are totally independent and can be performed in either order. Figure 2.4(a) shows an analog sampler

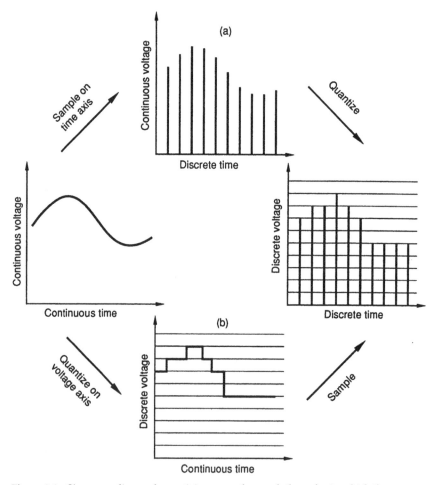

Figure 2.4 Since sampling and quantizing are orthogonal, the order in which they are performed is not important. At (a) sampling is performed first and the samples are quantized. This is common in audio convertors. At (b) the analog input is quantized into an asynchronous binary code. Sampling takes place when this code is latched on sampling clock edges. This approach is universal in video convertors.

preceding a quantizer, whereas (b) shows an asynchronous quantizer preceding a digital sampler. Ideally, both will give the same results; in practice each suffers from different deficiencies. Both approaches will be found in real equipment. In video convertors operating speed is a priority and the flash convertor is universally used. In the flash convertor the quantizing step is performed first and the quantized signal is subsequently sampled. In audio the sample accuracy is a priority and sampling first has the effect of freezing the signal voltage to allow time for an accurate quantizing process.

The independence of sampling and quantizing allows each to be discussed quite separately in some detail, prior to combining the processes for a full understanding of conversion.

2.8.1   Sampling and aliasing

Sampling is no more than periodic measurement, and it will be shown here that there is no theoretical need for sampling to be audible. Practical equipment may, of course, be less than ideal, but, given good engineering practice, the ideal may be approached quite closely.

Sampling must be precisely regular, because the subsequent process of timebase correction assumes a regular original process. The sampling process originates with a pulse train shown in Figure 2.5(a) to be of constant amplitude and period. The signal waveform amplitude-modulates the pulse train in much the same way as the carrier is modulated in an AM radio transmitter. One must be careful to avoid overmodulating the pulse train as shown in (b) and this is helped by applying a DC offset to the analog waveform so that, in audio, silence corresponds to a level half-way up the pulses as in (c). This approach will also be used in colour difference signals where blanking level will be shifted up to the midpoint of the scale. Clipping due to any excessive input level will then be symmetrical.

In the same way that AM radio produces sidebands or images above and below the carrier, sampling also produces sidebands; although the carrier is now a pulse train and has an infinite series of harmonics as can be seen in Figure 2.6(a). The sidebands in (b) repeat above and below each harmonic of the sampling rate.

The sampled signal can be returned to the continuous-time domain simply by passing it into a low-pass filter. This filter has a frequency response that prevents the images from passing, and only the baseband signal emerges, completely unchanged.

If an input is supplied having an excessive bandwidth for the sampling rate in use, the sidebands will overlap, and the result is aliasing, where

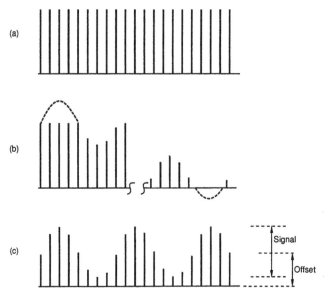

Figure 2.5   The sampling process requires a constant amplitude pulse train as shown at (a). This is amplitude modulated by the waveform to be sampled. If the input waveform has excessive amplitude or incorrect level, the pulse train clips as shown at (b). For an audio waveform, the greatest signal level is possible when an offset of half the pulse amplitude is used to centre the waveform as shown at (c).

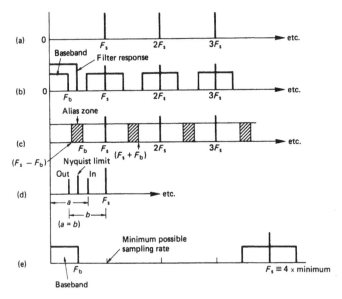

Figure 2.6   (a) Spectrum of sampling pulses. (b) Spectrum of samples. (c) Aliasing due to sideband overlap. (d) Beat-frequency production. (e) Four-times oversampling.

certain output frequencies are not the same as their input frequencies but become difference frequencies. It will be seen from (c) that aliasing occurs when the input frequency exceeds half the sampling rate, and this derives the most fundamental rule of sampling, first stated by Shannon in the West and at about the same time by Kotelnikov in Russia. This states that the sampling rate must be at least twice the highest input frequency.

In addition to the low-pass filter needed at the output to return to the continuous-time domain, a further low-pass filter is needed at the input to prevent aliasing. If input frequencies of more than half the sampling rate cannot reach the sampler, aliasing cannot occur.

Whilst aliasing has been described above in the frequency domain, it can equally be described in the time domain. In Figure 2.7(a) the sampling rate is obviously adequate to describe the waveform, but in (b) it is inadequate and aliasing has occurred.

Aliasing is commonly seen on television and in the cinema, owing to the relatively low frame rates used. With a frame rate of 24 Hz, a film camera will alias on any object changing at more than 12 Hz. Such objects include the spokes of stagecoach wheels, especially when being chased by Native Americans. When the spoke-passing frequency reaches 24 Hz the wheels appear to stop. Aliasing partly explains the inability of opinion polls to predict the results of elections. Aliasing does, however, have useful applications, including the stroboscope, which makes rotating machinery appear stationary, and the sampling oscilloscope, which can display periodic waveform of much greater frequency than the sweep speed of the tube normally allows.

In television systems the input image that falls on the camera sensor will be continuous in time, and continuous in two spatial dimensions corresponding to the height and width of the sensor. All three of these continuous dimensions will be sampled in a digital system. There is a direct connection between the concept of temporal sampling, where the input signal changes with respect to time at some frequency and is sampled at

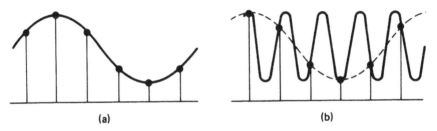

(a)   (b)

Figure 2.7 At (a) the sampling rate is adequate to reconstruct the original signal. At (b) the sampling rate is inadequate, and reconstruction produces the wrong waveform (dotted). Aliasing has taken place.

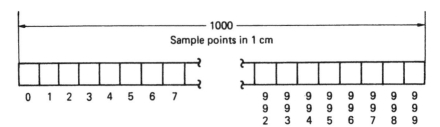

Figure 2.8 If the above spatial sampling arrangement of 1000 points per centimetre is scanned in 1 millisecond, the sampling rate will become 1 megahertz.

some other frequency, and spatial sampling, where an image changes a given number of times per unit distance and is sampled at some other number of times per unit distance. The connection between the two is the process of scanning. Temporal frequency can be obtained by multiplying spatial frequency by the speed of the scan. Figure 2.8 shows a hypothetical image sensor having 1000 discrete sensors across a width of 1 centimetre. The spatial sampling rate of this sensor is thus 1000 per centimetre. If the sensors are measured sequentially during a scan taking 1 millisecond to go across the 1 centimetre width, the result will be a temporal sampling rate of 1 MHz.

2.8.2 Reconstruction

If ideal low-pass anti-aliasing and anti-image filters are assumed, having a vertical cut-off slope at half the sampling rate, an ideal spectrum shown at Figure 2.9(a) is obtained. Figure 2.9(b) shows that the impulse response of a phase linear ideal low-pass filter is a sin x/x waveform in the time domain. Such a waveform passes through zero volts periodically. If the cut-off frequency of the filter is one-half of the sampling rate, the impulse passes through zero *at the sites of all other samples*. Thus at the output of such a filter, the voltage at the centre of a sample is due to that sample alone, since the value of *all* other samples is zero at that instant. In other words the continuous time output waveform must join up the tops of the input samples. In between the sample instants, the output of the filter is the sum of the contributions from many impulses, and the waveform smoothly joins the tops of the samples. If the time domain is being considered, the anti-image filter of the frequency domain can equally well be called the reconstruction filter. It is a consequence of the band-limiting of the original anti-aliasing filter that the filtered analog waveform could only travel between the sample points in one way. As the reconstruction filter has the same frequency response, the reconstructed output waveform must be identical to the original band-limited waveform prior to sampling.

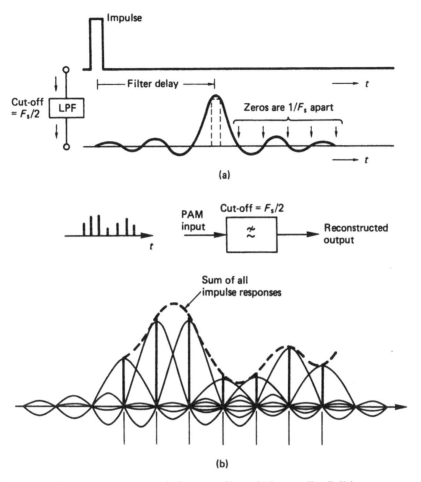

Figure 2.9 The impulse response of a low-pass filter which cuts off at $F_s/2$ has zeros at $1/F_s$ spacing which correspond to the position of adjacent samples, as shown at (b). The output will be a signal which has the value of each sample at the sample instant, but with smooth transitions from sample to sample.

It follows that sampling need not be audible. A rigorous mathematical proof of the above has been available since the 1930s, when PCM was invented, and can also be found in Betts[2].

2.8.3 Filter design

The ideal filter with a vertical 'brick-wall' cut-off slope is difficult to implement. As the slope tends to vertical, the delay caused by the filter goes to infinity: the quality is marvellous but you do not live to measure it. In practice, a filter with a finite slope is accepted as shown in Figure 2.10, and the sampling rate has to be raised a little to prevent aliasing. There is

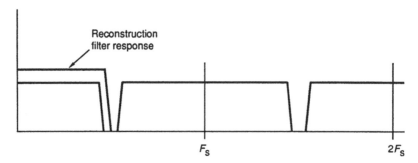

Figure 2.10 As filters with finite slope are needed in practical systems, the sampling rate is raised slightly beyond twice the highest frequency in the baseband.

no absolute factor by which the sampling rate must be raised; it depends upon the available filters.

It is not easy to specify such filters, particularly the amount of stopband rejection needed. The amount of aliasing resulting would depend on, among other things, the amount of out-of-band energy in the input signal. This is seldom a problem in video, but can warrant attention in audio where overspecified bandwidths are sometimes found. As a further complication, an out-of-band signal will be attenuated by the response of the anti-aliasing filter to that frequency, but the residual signal will then alias, and the reconstruction filter will reject it according to its attenuation at the new frequency to which it has aliased.

It could be argued that the reconstruction filter is unnecessary in audio, since all of the images are outside the range of human hearing. However, the slightest non-linearity in subsequent stages would result in gross intermodulation distortion. The possibility of damage to tweeters must also be considered. It would, however, be acceptable to bypass one of the filters involved in a copy from one digital machine to another via the analog domain, although a digital transfer is of course to be preferred. In video the filters are essential to constrain the bandwidth of the signal to the allowable broadcast channel width.

The nature of the filters used has a great bearing on the subjective quality of the system. Entire books have been written about analog filters, so they will only be treated briefly here. Figure 2.11 shows the terminology to be used to describe the common elliptic low-pass filter. These filters are popular because they can be realized with fewer components than other filters of similar response. It is a characteristic of these elliptic filters that there are ripples in the passband and stopband. In much equipment the anti-aliasing filter and the reconstruction filter will have the same specification, so that the passband ripple is doubled with a corresponding increase in dispersion. Sometimes slightly different filters are used to reduce the effect.

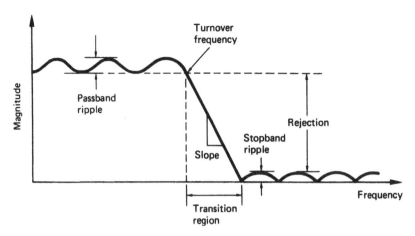

Figure 2.11   The important features and terminology of low-pass filters used for anti-aliasing and reconstruction.

It is difficult to produce an analog filter with low distortion. Passive filters using inductors suffer non-linearity at high levels due to the *B/H* curve of the cores. Active filters can simulate linear inductors using op-amp techniques, but they tend to suffer non-linearity at high frequencies where the falling open-loop gain reduces the effect of feedback. Active filters can also contribute noise, but this is not necessarily a bad thing in controlled amounts, since it can act as a dither source.

It is instructive to examine the phase response of such filters. Since a sharp cut-off is generally achieved by cascading many filter sections that cut at a similar frequency, the phase responses of these sections will accumulate. The phase may start to leave linearity at a fraction of the cut-off frequency and by the time this is reached, the phase may have completed several revolutions. Meyer[3] suggests that these phase errors are audible and that equalization is necessary. In video, phase linearity is essential as otherwise the different frequency components of a contrast step are smeared across the screen. An advantage of linear phase filters is that ringing is minimized, and there is less possibility of clipping on transients.

It is possible to construct a ripple-free phase-linear filter with the required stopband rejection[4,5], but the design effort and component complexity result in expense, and the filter might drift out of specification as components age. The effort may be more effectively directed towards avoiding the need for such a filter. Much effort can be saved in analog filter design by using oversampling. As shown in Figure 2.12 a high sampling rate produces a large spectral gap between the baseband and the first lower sideband. The anti-aliasing and reconstruction filters need only have a gentle roll-off, causing minimum disturbance to phase linearity in the baseband, and the

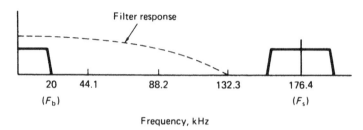

Figure 2.12   In this 4-times oversampling system, the large separation between baseband and sidebands allows a gentle roll-off reconstruction filter to be used.

Butterworth configuration, which does not have ripple or dispersion, can be used. The penalty of oversampling is that an excessive data rate results. It is necessary to reduce the rate using a digital low-pass filter (LPF). Digital filters can be made perfectly phase linear and, using LSI, can be inexpensive to construct. The superiority of oversampling convertors means that they have become universal in audio and appear set to do so in video.

2.8.4   Sampling clock jitter

The instants at which samples are taken in an A/D convertor (ADC) and the instants at which D/A convertors (DACs) make conversions must be evenly spaced; otherwise unwanted signals can be added to the waveform. Figure 2.13(a) shows the effect of sampling clock jitter on a sloping waveform. Samples are taken at the wrong times. When these samples have passed through a system, the timebase correction stage prior to the DAC will remove the jitter, and the result is shown in Figure 2.13(b). The magnitude of the unwanted signal is proportional to the slope of the signal waveform and so increases with frequency. The nature of the unwanted signal depends on the spectrum of the jitter. If the jitter is random, the effect is noise-like and relatively benign unless the amplitude is excessive. Clock jitter is, however, not necessarily random. Figure 2.14 shows that one source of clock jitter is crosstalk on the clock signal. The unwanted additional signal changes the time at which the sloping clock signal appears to cross the threshold voltage of the clock receiver. The threshold itself may be changed by ripple on the clock receiver power supply. There is no reason why these effects should be random; they may be periodic and potentially discernible.

The allowable jitter is measured in picoseconds in both audio and video signals, as shown in Figure 2.13, and clearly steps must be taken to eliminate it by design. Convertor clocks must be generated from clean power supplies that are well decoupled from the power used by the logic because a convertor clock must have a good signal-to-noise ratio. If an

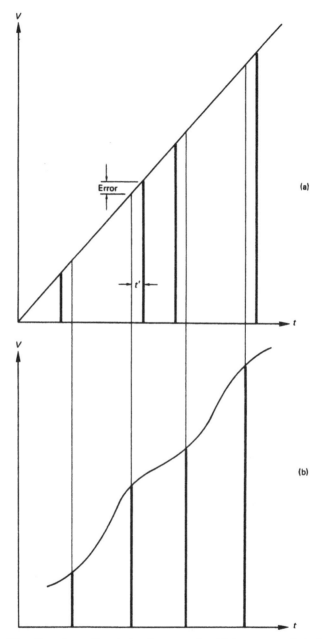

Figure 2.13   The effect of sampling timing jitter on noise. (a) A ramp sampled with jitter has an error proportional to the slope. (b) When the jitter is removed by later circuits, the error appears as noise added to samples. The superimposition of jitter may also be considered as a modulation process.

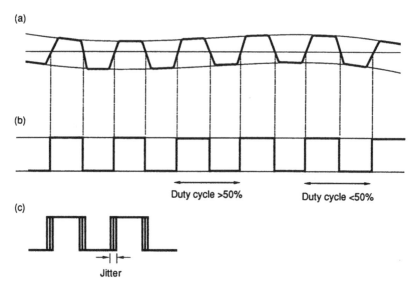

Figure 2.14 Crosstalk in transmission can result in unwanted signals being added to the clock waveform. It can be seen here that a low-frequency interference signal affects the slicing of the clock and causes a periodic jitter.

external clock is used, it cannot be used directly, but must be fed through a well-damped phase-locked loop that will filter out the jitter. The external clock signal is sometimes fed into the clean circuitry using an optical coupler to improve isolation.

Although it has been documented for many years, attention to control of clock jitter is not as great in actual audio hardware as it might be. It accounts for much of the slight audible differences between convertors reproducing the same data. A well-engineered convertor should substantially reject jitter on an external clock and should sound the same when reproducing the same data irrespective of the source of the data. A remote convertor which sounds different when reproducing, for example, the same Compact Disc via the digital outputs of a variety of CD players is simply not well engineered and should be rejected. Similarly if the effect of changing the type of digital cable feeding the convertor can be heard, the unit is a dud. Unfortunately many consumer external DACs fall into this category, as the steps outlined above have not been taken.

Jitter tends to be less noticeable on digital video signals and is generally not an issue until it becomes great enough to cause data errors.

2.8.5 Aperture effect

The reconstruction process of Figure 2.9 only operates exactly as shown if the impulses are of negligible duration. In many convertors this is not the case, and many keep the analog output constant until a different sample value is input and produces a waveform more like a staircase than a pulse train. In this case the pulses have effectively been extended in width to become equal to the sample period. This is known as a zero-order hold system and has a 100% aperture ratio. Note that the aperture effect is not apparent in a track-hold system; the holding period is for the convenience of the quantizer which outputs a value corresponding to the input voltage at the instant hold mode was entered.

Whereas pulses of negligible width have a uniform spectrum, which is flat within the audio band, pulses of 100% aperture have a sin x/x spectrum shown in Figure 2.15. The frequency response falls to a null at the sampling rate, and as a result is about 4 dB down at the edge of the baseband. If the pulse width is stable, the reduction of high frequencies is constant and predictable, and an appropriate equalization circuit can render the overall response flat once more. An alternative is to use resampling whose effect is shown in Figure 2.16. Resampling passes the zero-order-hold waveform through a further synchronous sampling stage consisting of an analog switch that closes briefly in the centre of each sample period. The output of the switch will be pulses that are narrower than the original. If, for example, the aperture ratio is reduced to 50% of the sample period, the first response null is now at twice the sampling rate, and the loss at the edge of the audio band is reduced. As the figure shows, the frequency response becomes flatter as the aperture ratio falls. The process should not be carried too far, as with very small aperture ratios there is little energy in the pulses and noise can be a problem. A practical limit is around 12.5% where the frequency response is virtually ideal.

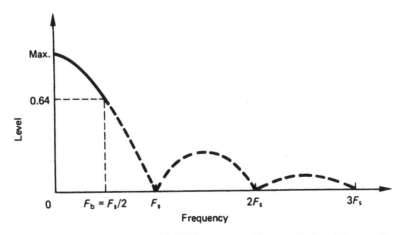

Figure 2.15 Frequency response with 100% aperture nulls at multiples of the sampling rate. The area of interest is up to half the sampling rate.

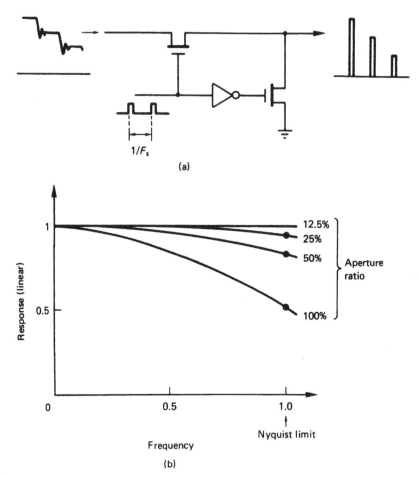

Figure 2.16 (a) A resampling circuit eliminates transients and reduces aperture ratio. (b) Response of various aperture ratios.

The aperture effect will show up in many aspects of television. Lenses have finite modulation transfer functions, such that a very small object becomes spread in the image. The image sensor will also have a finite aperture function. In tube cameras, the beam will have a finite radius, and will not necessarily have a uniform energy distribution across its diameter. In CCD cameras, the sensor is split into elements that may almost touch in some cases. The element integrates light falling on its surface, and so will have a rectangular aperture. In both cases there will be a roll-off of higher spatial frequencies.

It is highly desirable to prevent spatial aliasing, since the result is visually irritating. In tube cameras the aliasing will be in the vertical dimension only, since the horizontal dimension is continuously scanned. Such cameras seldom attempt to prevent vertical aliasing. CCD sensors can, however, alias in both horizontal and vertical dimensions, and so an anti-aliasing optical filter is generally fitted between the lens and the sensor. This takes the form of a plate that diffuses the image formed by the lens. Such a device can never have a sharp cut-off nor will the aperture be rectangular. The aperture of the anti-aliasing plate is in series with the aperture effect of the CCD elements, and the combination of the two effectively prevents spatial aliasing, and generally gives a good balance between horizontal and vertical resolution, allowing the picture a natural appearance.

Conventional tube cameras generally have better horizontal resolution, and produce vertical aliasing which has a similar spatial frequency to real picture information, but which lacks realism. With a conventional approach, there are effectively two choices. If aliasing is permitted, the theoretical information rate of the system can be approached. If aliasing is prevented, realizable anti-aliasing filters cannot sharp cut, and the information conveyed is below system capacity.

These considerations also apply at the television display. The display must filter out spatial frequencies above one-half the sampling rate. In a conventional CRT this means that an optical filter should be fitted in front of the screen to render the raster invisible. Again the aperture of a simply realizable filter would attenuate too much of the wanted spectrum, and so the technique is not used. The technique of spot wobble was most effective in reducing raster visibility in monochrome television sets without affecting horizontal resolution, and its neglect remains a mystery.

As noted, in conventional tube cameras and CRTs the horizontal dimension is continuous, whereas the vertical dimension is sampled. The aperture effect means that the vertical resolution in real systems will be less than sampling theory permits, and to obtain equal horizontal and vertical resolutions a greater number of lines are necessary. The magnitude of the increase is described by the so-called Kell factors[6], although the term factor is a misnomer since it can have a range of values depending on the apertures in use and the methods used to measure resolution[7]. In digital video, sampling takes place in horizontal and vertical dimensions, and the Kell parameter becomes unnecessary. The outputs of digital systems will, however, be displayed on raster scan CRTs, and the Kell parameter of the display will then be effectively in series with the other system constraints.

2.8.6   Choice of audio sampling rate

The Nyquist criterion is only the beginning of the process that must be followed to arrive at a suitable sampling rate. The slope of available filters will compel designers to raise the sampling rate above the theoretical Nyquist rate. For consumer products, the lower the sampling rate the better, since the cost of the medium is directly proportional to the sampling rate: thus sampling rates near to twice 20 kHz are to be expected. For professional products, there is a need to operate at variable speed for pitch correction. When the speed of a digital recorder is reduced, the offtape sampling rate falls, and Figure 2.17 shows that with a minimal sampling rate the first image frequency can become low enough to pass the reconstruction filter. If the sampling frequency is raised without changing

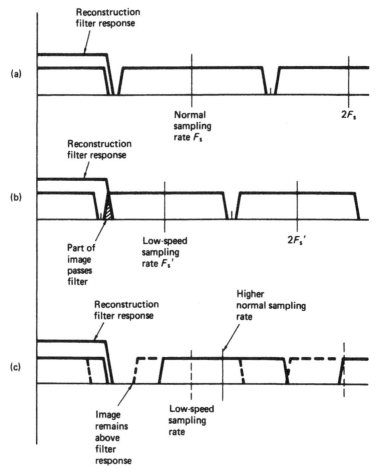

Figure 2.17 At normal speed, the reconstruction filter correctly prevents images entering the baseband, as at (a). (b) When speed is reduced, the sampling rate falls, and a fixed filter will allow part of the lower sideband of the sampling frequency to pass. (c) If the sampling rate of the machine is raised, but the filter characteristics remain the same, the problem can be avoided.

the response of the filters, the speed can be reduced without this problem. It follows that variable-speed recorders must use a higher sampling rate.

In the early days of digital audio research, the necessary bandwidth of about 1 megabit per second per audio channel was difficult to store. Disk drives had the bandwidth but not the capacity for long recording time, so attention turned to video recorders. These were adapted to store audio samples by creating a pseudo-video waveform that could convey binary as black and white levels[8]. The sampling rate of such a system is constrained to relate simply to the field rate and field structure of the television standard used, so that an integer number of samples can be stored on each usable TV line in the field. Such a recording can be made on a monochrome recorder, and these recordings are made in two standards, 525 lines at 60 Hz and 625 lines at 50 Hz. Thus it is possible to find a frequency which is a common multiple of the two and also suitable for use as a sampling rate.

The allowable sampling rates in a pseudo-video system can be deduced by multiplying the field rate by the number of active lines in a field (blanked lines cannot be used) and again by the number of samples in a line. By careful choice of parameters it is possible to use either 525/60 or 625/50 video with a sampling rate of 44.1 kHz.

In 60 Hz video, there are 35 blanked lines, leaving 490 lines per frame, or 245 lines per field for samples. If three samples are stored per line, the sampling rate becomes $60 \times 245 \times 3 = 44.1$ kHz.

In 50 Hz video, there are 37 lines of blanking, leaving 588 active lines per frame, or 294 per field, so the same sampling rate is given by $50 \times 294 \times 3 = 44.1$ kHz.

The sampling rate of 44.1 kHz came to be that of the Compact Disc. Even though CD has no video circuitry, the equipment used to make CD masters was originally video based and determined the sampling rate.

For landlines to FM stereo broadcast transmitters having a 15 kHz audio bandwidth, the sampling rate of 32 kHz is more than adequate, and has been in use for some time in the United Kingdom and Japan. This frequency is also in use in the NICAM 728 stereo TV sound system. The professional sampling rate of 48 kHz was proposed as having a simple relationship to 32 kHz, being far enough above 40 kHz for variable-speed operation, and having a simple relationship with PAL video timing which would allow digital video recorders to store the convenient number of 960 audio samples per video field. This is the sampling rate used by all of the professional DVTR formats[9]. The field rate offset of NTSC does not easily relate to any of the above sampling rates, and requires special handling that will be discussed further in this book.

Although in a perfect world the adoption of a single sampling rate might have had virtues, for practical and economic reasons digital audio now has essentially three rates to support: 32 kHz for broadcast, 44.1 kHz for CD, and 48 kHz for professional use[10]. Variations of the DAT format will support all of these rates, although the 32 kHz version is uncommon.

Recently there have been suggestions that extremely high sampling rates such as 96 and even 192 kHz are necessary for audio. Repeatable experiments to verify that this is the case seem to be elusive. It seems unlikely that all the world's psychoacoustic researchers would have so seriously underestimated the bandwidth of human hearing. It may simply be that the cost of electronic devices and storage media have fallen to the extent that there is little penalty in adopting such techniques, even if greater benefit would be gained by attending to areas in which progress would truly be beneficial, such as microphones and loudspeakers.

2.8.7 Choice of video sampling rate

Component or colour difference signals are used primarily for post-production work where quality and flexibility are paramount. In a digital colour difference system, the analog input video will be converted to the digital domain and will generally remain there whilst being operated on in digital effects units. The finished work will then be returned to the analog domain for transmission. The number of conversions is relatively few.

In contrast, composite digital video recorders are designed to work in an existing analog environment and so the number of conversions a signal passes through will be greater. It follows that composite will require a higher sampling rate than component so that filters with a more gentle slope can be used to reduce the build-up of response ripples.

In colour difference working, the important requirement is for image manipulation in the digital domain. This is facilitated by a sampling rate that is a multiple of line rate because then there is a whole number of samples in a line and samples are always in the same position along the line and can form neat columns. A practical difficulty is that the line period of the 525 and 625 systems is slightly different. The problem was overcome by the use of a sampling clock which is an integer multiple of both line rates. For standard definition working, the rate of 13.5 MHz is sufficient for the requirements of sampling theory, yet allows a whole number of samples in both line standards with a common clock frequency.

The colour difference signals have only half the bandwidth of the luminance and so can be sampled at one-half the luminance rate, i.e. 6.75 MHz. Extended and high definition systems require proportionately higher rates.

In composite video, the most likely process to be undertaken in the digital domain is decoding to components. This is performed, for example, in video recorders running in slow motion. Separation of luminance and chroma in the digital domain is simpler if the sampling rate is a multiple of the subcarrier frequency. Whilst three times the frequency of subcarrier is adequate from a sampling theory standpoint, this does require relatively steep filters, so as a practical matter four times subcarrier is used.

In high definition systems, clearly the sampling rate must rise. Doubling the resolution requires twice as many lines and twice as many pixels in each line, so the sampling rate must be quadrupled. However, a further factor is the adoption of 16:9 aspect ratio instead of 4:3 so that the rate becomes yet higher as each line has more than twice as many pixels. In some cases 74.25 MHz is used. Eventually it became clear that the proliferation of frame rates and image sizes would make it impossible to interface. The solution for HD was to adopt what is essentially a fixed clock rate interface that can support a variety of actual data rates by using variable amounts of blanking. Thus the actual video sampling rate can be whatever is appropriate for the picture size and rate to give square pixels.

2.8.8 Quantizing

Quantizing is the process of expressing some infinitely variable quantity by discrete or stepped values. Quantizing turns up in a remarkable number of everyday guises. Figure 2.18 shows that an inclined ramp allows infinitely variable height, whereas on a stepladder only discrete height is possible. A stepladder quantizes height. When accountants round off sums of money to the nearest pound or dollar they are quantizing.

In audio and video the values to be quantized are infinitely variable voltages from an analog source. Strict quantizing is a process restricted to the signal amplitude domain only. For the purpose of studying the quantizing of a single sample, time is assumed to stand still. This is achieved in practice either by the use of a track-hold circuit or the adoption of a quantizer technology that operates before the sampling stage.

Figure 2.19(a) shows that the process of quantizing divides the signal voltage range into quantizing intervals Q. In applications such as telephony these may be of differing size, but for digital audio and video the quantizing

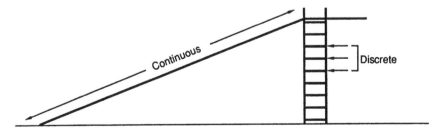

Figure 2.18 An analog parameter is continuous whereas a quantized parameter is restricted to certain values. Here the sloping side of a ramp can be used to obtain any height whereas a ladder only allows discrete heights.

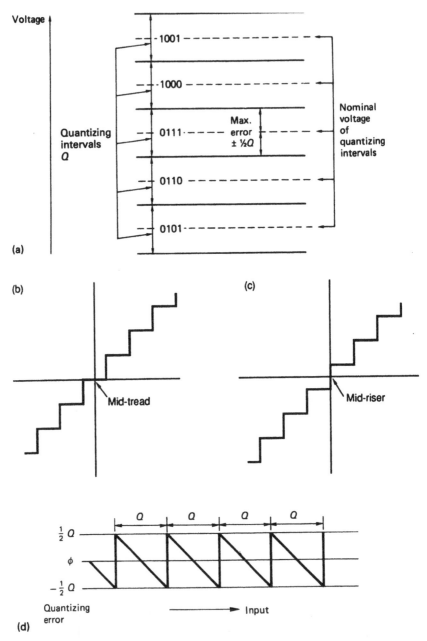

Figure 2.19 Quantizing assigns discrete numbers to variable voltages. All voltages within the same quantizing interval are assigned the same number which causes the DAC to produce the voltage at the centre of the intervals shown by the dashed lines at (a). This is the characteristic of the mid-tread quantizer shown at (b). An alternative system is the mid-riser system shown at (c). Hero 0 volts analog falls between two codes and there is no code for zero. Such quantizing cannot be used prior to signal processing because the number is no longer proportional to the voltage. Quantizing error cannot exceed ±0.5Q as shown at (d).

intervals are made as identical as possible. If this is done, the binary numbers which result are truly proportional to the original analog voltage, and the digital equivalents of mixing and gain changing can be performed by adding and multiplying sample values. If the quantizing intervals are unequal this cannot be done. When all quantizing intervals are the same, the term uniform quantizing is used. The term linear quantizing will be found, but this is, like military intelligence, a contradiction in terms.

The term LSB (Least Significant Bit) will also be found in place of quantizing interval in some treatments, but this is a poor term because quantizing is not always used to create binary values and because a bit can only have two values. In studying quantizing we wish to discuss values smaller than a quantizing interval, but a fraction of an LSB is a contradiction in terms.

Whatever the exact voltage of the input signal, the quantizer will determine the quantizing interval in which it lies. In what may be considered a separate step, the quantizing interval is then allocated a code value that is typically some form of binary number. The information sent is the number of the quantizing interval in which the input voltage lay. Exactly where that voltage lay within the interval is not conveyed, and this mechanism puts a limit on the accuracy of the quantizer. When the number of the quantizing interval is converted back to the analog domain, it will result in a voltage at the centre of the quantizing interval as this minimizes the magnitude of the error between input and output. The number range is limited by word length of the binary numbers used. In a 16-bit system commonly used for audio, 65 536 different quantizing intervals exist, whereas video systems typically have eight-bit systems having 256 quantizing intervals.

2.8.9 Quantizing error

It is possible to draw a transfer function for such an ideal quantizer followed by an ideal DAC, and this is shown in Figure 2.19(b). A transfer function is simply a graph of the output with respect to the input. When the term linearity is used, this generally means the straightness of the transfer function. Linearity is a goal in audio and video, yet it will be seen that an ideal quantizer is anything but.

Figure 2.19(b) shows that the transfer function is somewhat like a staircase, and the voltage corresponding to audio muting or video blanking is half way up a quantizing interval, or on the centre of a tread. This is the so-called mid-tread quantizer universally used in audio and video. Figure 2.19(c) shows the alternative mid-riser transfer function that causes difficulty because it does not have a code value at muting/blanking level and as a result the code value is not proportional to the signal voltage.

Quantizing causes a voltage error in the sample given by the difference between the actual staircase transfer function and the ideal straight line. This is shown in Figure 2.19(d) to be a sawtooth function periodic in Q. The amplitude cannot exceed $\pm 0.5Q$ unless the input is so large that clipping occurs.

In studying the transfer function it is better to avoid complicating matters with the aperture effect of the DAC. For this reason it is assumed here that output samples are of negligible duration. Then impulses from the DAC can be compared with the original analog waveform and the difference will be impulses representing the quantizing error waveform. As can be seen in Figure 2.20 the quantizing error waveform can be thought of as an unwanted signal that the quantizing process adds to the perfect original. As the transfer function is non-linear, ideal quantizing can cause distortion. As a result practical digital audio devices use non-ideal quantizers to achieve linearity. The quantizing error of an ideal quantizer is a complex function, and it has been researched in great depth[11]. It is not intended to go into

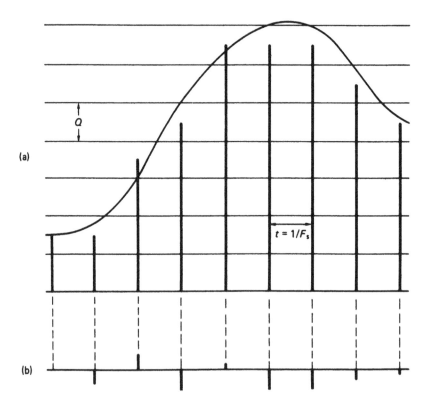

Figure 2.20   At (a) an arbitary signal is represented to finite accuracy by PAM needles whose peaks are at the centre of the quantizing intervals. The errors caused can be thought of as an unwanted signal (b) added to the original.

such depth here. The characteristics of an ideal quantizer will only be pursued far enough to convince the reader that they cannot be used.

As the magnitude of the quantizing error is limited, the effect can be minimized by the use of a larger signal. This will require more quantizing intervals and more bits to express them. The number of quantizing intervals multiplied by their size gives the quantizing range of the convertor. A signal outside the range will be clipped. Clearly if clipping is avoided, the larger the signal the less will be the effect of the quantizing error.

Consider first the case where the input signal exercises the whole quantizing range and has a complex waveform. In audio this might be orchestral music; in video a bright, detailed contrasting scene. In these cases successive samples will have widely varying numerical values and the quantizing error on a given sample will be independent of that on others.

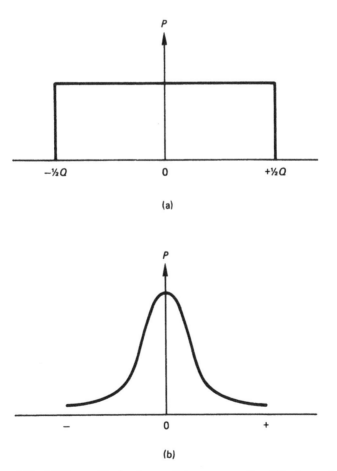

Figure 2.21 (a) The amplitude of a quantizing error needle will be from $-0.5Q$ to $+0.5Q$ with equal probability. (b) White noise in analog circuits generally has Gaussian amplitude distribution.

In this case the size of the quantizing error will be distributed with equal probability between the limits. Figure 2.21(a) shows the resultant uniform probability density. In this case the unwanted signal added by quantizing is an additive broadband noise uncorrelated with the signal, and it is appropriate in this case to call it quantizing noise. This is not quite the same as thermal noise which has a Gaussian probability shown in Figure 2.21(b). The subjective difference is slight. Treatments that then assume that quantizing error *is always* noise give results at variance with reality. Such approaches only work if the probability density of the quantizing error is uniform. Unfortunately at low levels, and particularly with pure or simple waveforms, this is simply not true.

At low levels, quantizing error ceases to be random, and becomes a function of the input waveform and the quantizing structure. Once an unwanted signal becomes a deterministic function of the wanted signal, it has to be classed as a distortion rather than a noise. We predicted a distortion because of the non-linearity or staircase nature of the transfer function. With a large signal, there are so many steps involved that we must stand well back, and a staircase with many steps appears to be a slope. With a small signal there are few steps and they can no longer be ignored.

The non-linearity of the transfer function results in distortion, which produces harmonics. Unfortunately these harmonics are generated *after* the anti-aliasing filter, and so any which exceed half the sampling rate will alias. Figure 2.22 shows how this results in anharmonic distortion in audio. These anharmonics result in spurious tones known as bird-singing. When the sampling rate is a multiple of the input frequency the result is harmonic distortion. This is shown in Figure 2.23. Where more

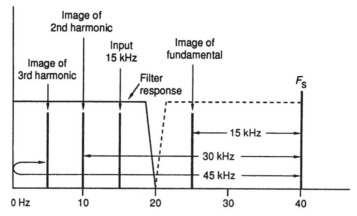

Figure 2.22 Quantizing produces distortion *after* the anti-aliasing filter, thus the distortion products will fold back to produce anharmonics in the audio band. Here the fundamental of 15 kHz produces 2nd and 3rd harmonic distortion at 30 and 45 kHz. This results in aliased products at 40–30 = 10 kHz and 40–45 = (−)5 kHz.

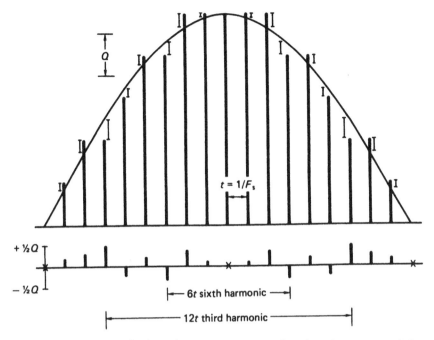

Figure 2.23 Mathematically derived quantizing error waveform for a sine wave sampled at a multiple of itself. The numerous autocorrelations between quantizing errors show that there are harmonics of the signal in the error, and that the error is not random, but deterministic.

than one frequency is present in the input, intermodulation distortion occurs, which is known as granulation.

As the input signal is further reduced in level, it may remain within one quantizing interval. The output will be silent because the signal is now the quantizing error. In this condition, low-frequency signals such as air-conditioning rumble can shift the input in and out of a quantizing interval so that the quantizing distortion comes and goes, resulting in noise modulation.

In video, quantizing error results in visible contouring on low-key scenes or flat fields. Continuous change of brightness across the screen is replaced by a series of sudden steps between areas of constant brightness.

Needless to say any one of the above effects would prevent the use of an ideal quantizer for high quality work. There is little point in studying the adverse effects further as they can be eliminated completely in practical equipment by the use of dither.

2.8.10 Dither

At high signal level, quantizing error is effectively noise. As the level falls, the quantizing error of an ideal quantizer becomes more strongly correlated with the signal and the result is distortion. If the quantizing error can be decorrelated from the input in some way, the system can remain linear. Dither performs the job of decorrelation by making the action of the quantizer unpredictable.

The first documented use of dither was in picture coding[12]. In this system, the noise added prior to quantizing was subtracted after reconversion to analog. This is known as subtractive dither. Although subsequent subtraction has some slight advantages[11], it suffers from practical drawbacks, since the original noise waveform must accompany the samples or must be synchronously recreated at the DAC. This is virtually impossible in a system where the signal may have been edited. Practical systems use non-subtractive dither where the dither signal is added prior to quantization and no subsequent attempt is made to remove it. The introduction of dither inevitably causes a slight reduction in the signal-to-noise ratio attainable, but this reduction is a small price to pay for the elimination of non-linearity. As linearity is an essential requirement for digital audio and video, the use of dither is equally essential and there is little point in deriving a signal-to-noise ratio for an undithered system as this has no relevance to real applications. Instead, a study of dither will be used to derive the dither amplitude necessary for linearity and freedom from artefacts, and the signal-to-noise ratio available should follow from that.

The ideal (noiseless) quantizer of Figure 2.19 has fixed quantizing intervals and must always produce the same quantizing error from the same signal. In Figure 2.24 it can be seen that an ideal quantizer can be dithered by linearly adding a controlled level of noise either to the input signal or to the reference voltage used to derive the quantizing intervals. There are several ways of considering how dither works, all of which are valid. The addition of dither means that successive samples effectively find the quantizing intervals in different places on the voltage scale. The quantizing error becomes a function of the dither, rather than just a function of the input signal. The quantizing error is not eliminated, but the subjectively unacceptable distortion is converted into broadband noise that is more benign.

That noise may be also frequency shaped to take account of the frequency shaping of the ear and eye responses.

An alternative way of looking at dither is to consider the situation where a low level input signal is changing slowly within a quantizing interval. Without dither, the same numerical code results, and the variations within the interval are lost. Dither has the effect of forcing the quantizer to switch between two or more states. The higher the voltage of the input signal within the interval, the more probable it becomes that the output code will take on a higher value. The lower the input voltage within the interval, the more probable it is that the output code will take the lower value. The dither has resulted in a form of duty cycle modulation, and the resolution

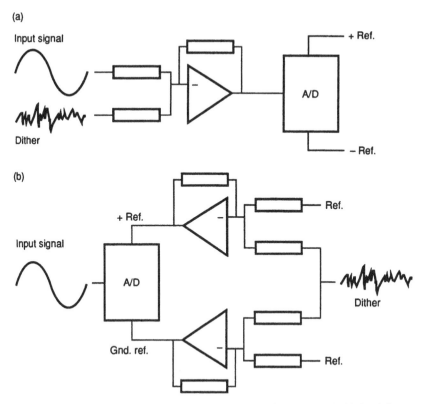

Figure 2.24 Dither can be applied to a quantizer in one of two ways. At (a) the dither is linearly added to the analog input signal whereas at (b) it is added to the reference voltages of the quantizer.

of the system has been extended indefinitely instead of being limited by the size of the steps.

Dither can also be understood by considering the effect it has on the transfer function of the quantizer. This is normally a perfect staircase, but in the presence of dither it is smeared horizontally until, when the amplitude has increased sufficiently, the average transfer function becomes straight.

The characteristics of the noise used are rather important for optimal performance, although many suboptimal but nevertheless effective systems are in use. The main parameters of interest are the peak-to-peak amplitude, and the probability distribution of the amplitude. Triangular probability works best and this can be obtained by summing the output of two uniform probability processes.

The use of dither invalidates the conventional calculations of signal-to-noise ratio available for a given word length. This is of little consequence as the rule of thumb that multiplying the number of bits in the word length by 6 dB gives the S/N ratio will be close enough for all practical purposes.

The technique can also be used to convert signals quantized with more intervals to a lower number of intervals.

It has only been possible to introduce the principles of conversion of audio and video signals here. For more details of the operation of convertors the reader is referred elsewhere[13,14].

2.9 Binary codes for audio

For audio use, the prime purpose of binary numbers is to express the values of the samples representing the original analog sound-pressure waveform. There will be a fixed number of bits in the sample, which determines the number range. In a 16-bit code there are 65 536 different numbers. Each number represents a different analog signal voltage, and care must be taken during conversion to ensure that the signal does not go outside the convertor range, or it will be clipped. In Figure 2.25, it will be seen that in a unipolar system, the number range goes from 0000 hex, which represents the largest negative voltage, through 7FFF hex, which represents the smallest negative voltage, through 8000 hex, which represents the smallest positive voltage, to FFFF hex, which represents the largest positive voltage. Effectively the number range of the convertor has been shifted so that positive and negative voltages in a bipolar signal may be expressed with a positive-only binary number. This approach is called

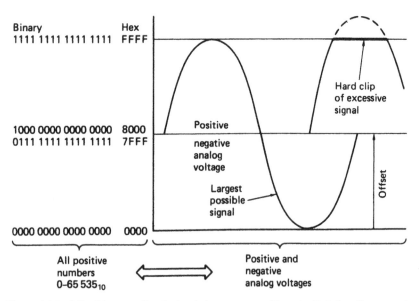

Figure 2.25 Offset binary coding is simple but causes problems in digital audio processing. It is seldom used.

offset binary, and is perfectly acceptable where the signal has been digitized only for recording or transmission from one place to another, after which it will be converted back to analog. Under these conditions it is not necessary for the quantizing steps to be uniform, provided both ADC and DAC are constructed to the same standard. In practice, it is the requirements of signal processing in the digital domain that make both non-uniform quantizing and offset binary unsuitable.

Figure 2.26 shows an audio signal voltage is referred to midrange. The level of the signal is measured by how far the waveform deviates from midrange, and attenuation, gain and mixing all take place around midrange. It is necessary to add sample values from two or more different sources to perform the mixing function, and adding circuits assumes that all bits represent the same quantizing interval so that the sum of two sample values will represent the sum of the two original analog voltages. In non-uniform quantizing this is not the case, and such signals cannot readily be processed. If two offset binary sample streams are added together in an attempt to perform digital mixing, the result will be an offset that may lead to an overflow. Similarly, if an attempt is made to attenuate by, say, 6 dB by dividing all of the sample values by two, Figure 2.27 shows that a further offset results. The problem is that offset binary is referred to one end of the range. What is needed is a numbering system that operates symmetrically about the centre of the range.

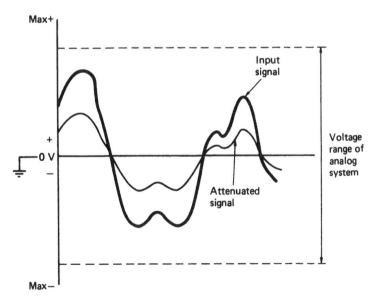

Figure 2.26 Attenuation of an audio signal takes place with respect to midrange.

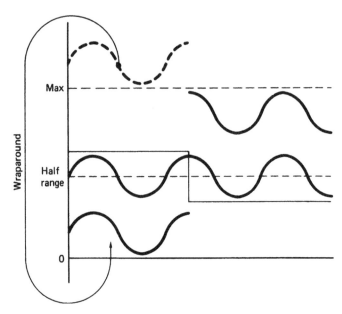

Figure 2.27   If two pure binary data streams are added to simulate mixing, offset or overflow will result.

In the two's complement system, which has this property, the upper half of the pure binary number range has been defined to represent negative quantities. If a pure binary counter is constantly incremented and allowed to overflow, it will produce all the numbers in the range permitted by the number of available bits, and these are shown for a four-bit example drawn around the circle in Figure 2.28. In two's complement, however, the number range this represents does not start at zero, but starts on the opposite side of the circle. Zero is midrange, and all numbers with the most significant bit set are considered negative. This system allows two sample values to be added, where the result is referred to the system midrange; this is analogous to adding analog signals in an operational amplifier. A further asset of two's complement notation is that binary subtraction can be performed using only adding logic. The two's complement is added to perform a subtraction. This permits a significant reduction of hardware complexity, since only carry logic is necessary and no borrow mechanism need be supported.

For these reasons, two's complement notation is in virtually universal use in digital audio processing, and is accordingly adopted by all the major digital recording formats, and consequently it is specified for the AES/EBU digital audio interface.

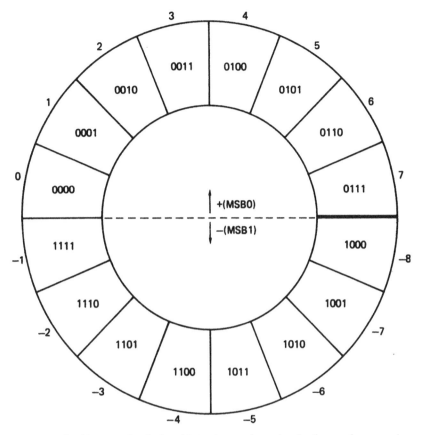

Figure 2.28 In this example of a four-bit two's complement code, the number range is from −8 to +7. Note that the MSB determines polarity.

Fortunately the process of conversion to two's complement is simple. The signed binary number to be expressed as a negative number is written down with leading zeros if necessary to occupy the word length of the system. All bits are then inverted, to form the one's complement, and one is added. To return to signed binary, if the most significant bit of the two's complement number is set to zero, no action needs to be taken. If the most significant bit is one, the sign is negative, all bits are inverted, and one is added. Figure 2.29 shows some examples of conversion to and from two's complement, and illustrates how the addition of two's complement samples simulates the mixing process.

Reference to Figure 2.28 will show that inverting the MSB results in a jump to the diametrically opposite code that is the equivalent of an offset of half full scale. Thus it is possible to obtain two's complement coded data by adding an offset of half full scale to the analog input of a convertor and then inverting the MSB as shown in Figure 2.30.

(a) **Conversion to two s complement from binary**
Positive numbers: add leading zeros to determine sign bit.
Example: $101_2 = 5_{10} = 0101_{2C}$

Negative numbers: add leading zeros to final
wordlength; invert all bits; add one.
Example 1: $11_2 = 3_{10} \rightarrow 0011 \rightarrow 1100 \rightarrow 1101_{2C} = -3$

add
leading   invert   add
zeros   1

Example 2: $100_2 = 4_{10} \rightarrow 0100 \rightarrow 1011 \rightarrow 1100_{2C} = -4$

(b) **Conversion to binary from two's complement**
If MSB = 1 (Negative Number), invert all bits; add one.

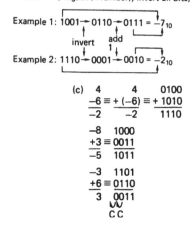

Example 1: $1001 \rightarrow 0110 \rightarrow 0111 = -7_{10}$

invert   add
1

Example 2: $1110 \rightarrow 0001 \rightarrow 0010 = -2_{10}$

(c)   4   4   0100
      $-6 \equiv + (-6) \equiv + 1010$
      ─── ───── ────
      -2 -2 1110

      -8 1000
      $+3 \equiv 0011$
      ───   ────
      -5 1011

      -3 1101
      $+6 \equiv 0110$
      ──   ────
       3   0011
                 C C

(d)

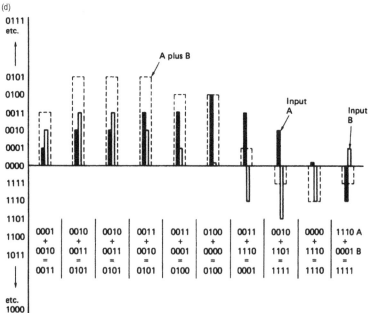

0001	0010	0010	0011	0011	0100	0011	0010	0000	1110 A
+	+	+	+	+	+	+	+	+	+
0010	0011	0011	0010	0001	0000	1110	1101	1110	0001 B
=	=	=	=	=	=	=	=	=	=
0011	0101	0101	0101	0100	0100	0001	1111	1110	1111

Figure 2.29 (a) Two's complement conversion from binary. (b) Binary conversion from two's complement. (c) Some examples of two's complement arithmetic. (d) Using two's complement arithmetic, single values from two waveforms are added together with respect to midrange to give a correct mixing function.

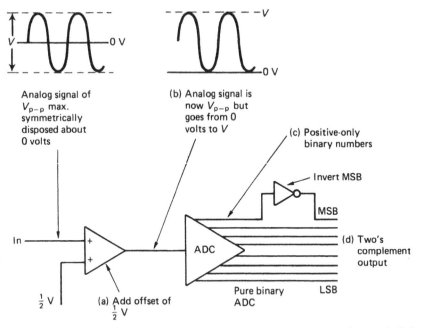

Figure 2.30 A two's complement ADC. At (a) an analog offset voltage equal to one-half the quantizing range is added to the bipolar analog signal in order to make it unipolar as at (b). The ADC produces positive only numbers at (c), but the MSB is then inverted at (d) to give a two's complement output.

2.10 Binary codes for video

There are a wide variety of waveform types encountered in video, and all of these can be successfully digitized. All that is needed is for the quantizing range to be a little greater than the useful voltage range of the waveform.

Composite video contains an embedded subcarrier, which has meaningful levels above white and below black. The quantizing range has to be extended to embrace the total possible excursion of luminance plus subcarrier. The resultant range is so close to the overall range of the signal that, in composite working, the whole signal, including syncs, is made to fit into the quantizing range as Figure 2.31 (b) shows. This is particularly useful in PAL because the sampling points are locked to subcarrier, and have a complex relationship to sync. Clearly black level no longer corresponds to digital zero. In eight-bit PAL it is 64_{10}. It is as if a constant of 64 had been added to every sample, and gives rise to the term offset binary.

In the luminance component signal, the useful waveform is unipolar and can only vary between black and white, since the syncs carry no information that cannot be recreated. Only the active line is coded and the quantizing range is optimized to suit the gamut of unblanked luminance as shown in Figure 2.31 (c). The same approach is used in HD component signals.

Chroma and colour difference signals are bipolar and can have values above and below blanking. In this they have the same characteristics as audio and require to be *processed* using two's complement coding as described in the previous section. However, two's complement is not used in standard digital video interfaces as the codes of all ones and all zeros are reserved for synchronizing and in two's complement these would appear in the centre of the quantizing range. Instead digital video interfaces use an offset of half full scale to shift blanking level to the centre of the scale as shown in Figure 2.31 (d). All ones and all zeros are then at the ends of the scale. Such an offset binary signal can easily be converted to two's complement by inverting the MSB as shown in Figure 2.30. If the quantizing range were set to exactly the video signal range or gamut, a slightly excessive gain somewhere in the analog input path would cause clipping. In practice the quantizing range will be a little greater than the nominal signal range and as a result even the unipolar luminance signal has an offset blanking level. A video interface format must specify the numerical values that result from reference analog levels so that all transmissions will result in identical signals at all receiving devices.

2.11 Requantizing and digital dither

The advanced ADC technology now available allows 20 or more bits of resolution to be obtained in audio. The situation then arises that an existing 16-bit device such as a digital recorder needs to be connected to the output of an ADC having greater word length. In a similar fashion digital video equipment has recently moved up from eight- to ten-bit working with the result that ten-bit signals are presented to eight-bit devices. In both cases the words need to be shortened in some way.

When a sample value is attenuated, the extra low-order bits that come into existence below the radix point preserve the resolution of the signal. The dither in the least significant bit(s) linearizes the system. The same word extension will occur in any process involving multiplication, such as mixing or digital filtering. It will subsequently be necessary to shorten the word length. Clearly the high-order bits cannot be discarded in two's complement as this would cause clipping of positive half cycles and a level shift on negative half cycles due to the loss of the sign bit. Low-order bits must be removed instead. Even if the original conversion was correctly dithered, the random element in the low-order bits will now be some way below the end of the intended word. If the word is simply truncated by discarding the unwanted low-order bits or rounded to the nearest integer the linearizing effect of the original dither will be lost. In audio the result will be low-level distortion; in video the result will be contouring.

Shortening the word length of a sample reduces the number of quantizing intervals available without changing the signal amplitude. As Figure 2.32 shows, the quantizing intervals become larger and the original signal is *requantized* with the new interval structure. This will introduce requantizing distortion having the same characteristics as quantizing distortion in an ADC. It then is obvious that when shortening the word length

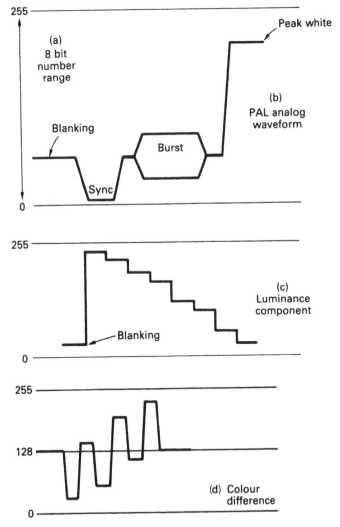

Figure 2.31 The unipolar quantizing range of an eight-bit pure binary system is shown at (a). The analog input must be shifted to fit into the quantizing range, as shown for PAL at (b). In component, sync pulses are not digitized, so the quantizing intervals can be smaller as at (c). An offset of half scale is used for colour difference signals (d).

of a 20-bit convertor to 16 bits, the four low-order bits must be removed in a way that displays the same overall quantizing structure as if the original convertor had been only of 16-bit word length. It will be seen from Figure 2.32 that truncation cannot be used because it does not meet the above requirement but results in signal dependent offsets because it always rounds in the same direction. Proper numerical rounding is essential in audio and video applications. Rounding in two's complement is a little more complex than in pure binary. Requantizing by numerical rounding accurately simulates analog quantizing to the new interval size. Unfortunately the 20-bit convertor will have a dither amplitude appropriate to quantizing intervals one-sixteenth the size of a 16-bit unit and the result will be highly non-linear.

In practice, the word length of samples must be shortened in such a way that the requantizing error is converted to noise rather than distortion. One technique that meets this requirement is to use digital dithering[15] prior to rounding. This is directly equivalent to the analog dithering in an ADC.

Digital dither is a pseudo-random sequence of numbers. If it is required to simulate the analog dither signal of Figure 2.24, then it is obvious that the noise must be bipolar so that it can have an average voltage of zero. Two's complement coding must be used for the dither values to obtain this characteristic.

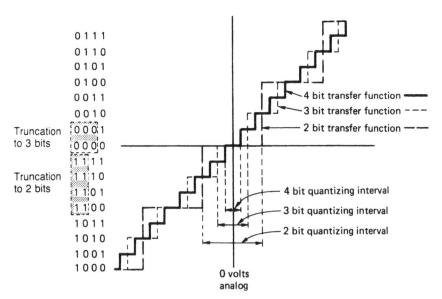

Figure 2.32 Shortening the word length of a sample reduces the number of codes which can describe the voltage of the waveform. This makes the quantizing steps bigger, hence the term requantizing. It can be seen that simple truncation or omission of the bits does not give analogous behaviour. Rounding is necessary to give the same result as if the larger steps had been used in the original conversion.

Figure 2.33 shows a simple digital dithering system (i.e. one without noise shaping) for shortening sample word length. The output of a two's complement pseudo-random sequence generator of appropriate word length is added to input samples prior to rounding. The most significant of the bits to be discarded is examined in order to determine whether the bits to be removed sum to more or less than half a quantizing interval. The dithered sample is either rounded down, that is, the unwanted bits are simply discarded, or rounded up, that is the unwanted bits are discarded but 1 is added to the value of the new short word. The rounding process is no longer deterministic because of the added dither that provides a linearizing random component.

If this process is compared with that of Figure 2.24 it will be seen that the principles of analog and digital dither are identical; the processes simply take place in different domains using numbers that are rounded or voltages that are quantized as appropriate. In fact quantization of an analog dithered waveform is identical to the hypothetical case of rounding after bipolar digital dither where the number of bits to be removed is infinite, and remains identical for practical purposes when as few as eight bits are to be removed. The probability density of the pseudo-random sequence is important. Vanderkooy and Lipshitz[15] found that uniform probability density produced noise modulation, in which the amplitude of the random component varies as a function of the amplitude of the samples. A triangular probability density function obtained by adding together two pseudo-random sequences eliminated the noise modulation to yield a signal-independent

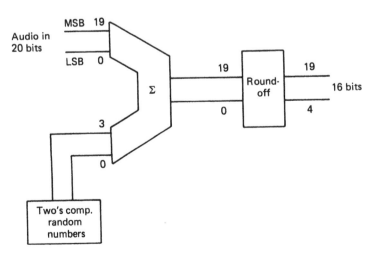

Figure 2.33 In a simple digital dithering system, two's complement values from a random number generator are added to low-order bits of the input. The dithered values are then rounded up or down according to the value of the bits to be removed. The dither linearizes the requantizing.

white-noise component in the least significant bit. It is vital that such steps are taken when sample word length is to be reduced. More recently, sequences yielding signal-independent noise with a weighted frequency component have been used to improve the subjective effects of the noise component leading to a perceived reduction in noise.

2.12 Introduction to compression

Compression, bit rate reduction and data reduction are all terms meaning basically the same thing in this context. In essence the same (or nearly the same) information is carried using a smaller quantity or rate of data. In transmission systems, compression allows a reduction in bandwidth and will generally result in a reduction in cost to make possible some process that would be impracticable without it. If a given bandwidth is available to an uncompressed signal, compression allows faster than real-time transmission in the same bandwidth. If a given bandwidth is available, compression allows a better quality signal in the same bandwidth.

Compression is summarized in Figure 2.34. It will be seen in (a) that the data rate is reduced at source by the compressor. The compressed data is then passed through a communication channel and returned to the original rate by the expander. The ratio between the source data rate and the channel data rate is called the compression factor. The term coding gain is also used. Sometimes a compressor and expander in series are referred to as a compander. The compressor may equally well be referred to as a coder and the expander a decoder in which case the tandem pair may be called a codec.

In audio and video compression, where the encoder is more complex than the decoder the system is said to be asymmetrical as in (b). The encoder needs to be algorithmic or adaptive whereas the decoder is 'dumb' and only carries out actions specified by the incoming bitstream. This is advantageous in applications such as broadcasting where the number of expensive complex encoders is small but the number of simple inexpensive decoders is large. In point-to-point applications the advantage of asymmetrical coding is not so great.

Although there are many different coding techniques, all of them fall into one or other of these categories. In lossless coding, the data from the expander is identical bit for bit with the original source data. Lossless coding is generally restricted to compression factors of around 2:1. A lossless coder cannot guarantee a particular compression factor and the channel used with it must be able to function with the variable output data rate.

In lossy coding data from the expander is not identical bit for bit with the source data and as a result comparison of the input with the output is bound

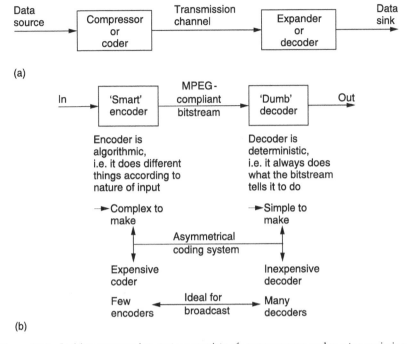

Figure 2.34 In (a) a compression system consists of compressor or coder, a transmission channel and a matching expander or decoder. The combination of coder and decoder is known as a codec. (b) MPEG is asymmetrical since the encoder is much more complex than the decoder.

to reveal a difference. Lossy codecs are not suitable for computer data, but are used in MPEG[16] as they allow greater compression factors than lossless codecs. Successful lossy codecs are those in which the errors are arranged so that a human viewer or listener finds them subjectively difficult to detect. Thus lossy codecs must be based on an understanding of psycho-acoustic and psycho-visual perception and are often called perceptive codes.

In perceptive coding, the greater the compression factor required, the more accurately must the human senses be modelled. Perceptive coders can be forced to operate at a fixed compression factor. This is convenient for practical transmission applications where a fixed data rate is easier to handle than a variable rate. Source data that results in poor compression factors on a given codec is described as difficult. The result of a fixed compression factor is that the subjective quality can vary with the 'difficulty' of the input material. Perceptive codecs should not be concatenated indiscriminately especially if they use different algorithms.

Although the adoption of digital techniques is recent, compression itself is as old as television. Figure 2.35 shows some of the compression techniques used in traditional television systems.

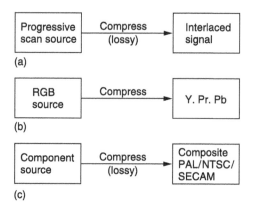

Figure 2.35 Compression is a old as television. (a) Interlace is a primitive way of halving the bandwidth. (b) Colour difference working invisibly reduces colour resolution. (c) Composite video transmits colour in the same bandwidth as monochrome.

One of the oldest techniques is interlace, which has been used in analog television from the very beginning as a primitive way of reducing bandwidth. Interlace is not without its problems, particularly in motion rendering. MPEG-2 supports interlace simply because legacy interlaced signals exist and there is a requirement to compress them. This should not be taken to mean that it is a good idea.

The generation of colour difference signals from RGB in video represents an application of perceptive coding. The human visual system (HVS) sees no change in quality although the bandwidth of the colour difference signals is reduced. This is because human perception of detail in colour changes is much less than in brightness changes. This approach is sensibly retained in MPEG.

Composite video systems such as PAL, NTSC and SECAM are all analog compression schemes that embed a subcarrier in the luminance signal so that colour pictures are available in the same bandwidth as monochrome. In comparison with a progressive scan RGB picture, interlaced composite video has a compression factor of 6:1.

In a sense MPEG-2 can be considered to be a modern digital equivalent of analog composite video as it has most of the same attributes. For example, the eight-field sequence of PAL subcarrier that makes editing difficult has its equivalent in the GOP (group of pictures) of MPEG.

In a PCM digital system the bit rate is the product of the sampling rate and the number of bits in each sample and this is generally constant. Nevertheless the information rate of a real signal varies. In all real signals, part of the signal is obvious from what has gone before or what may come later and a suitable receiver can predict that part so that only the true information actually has to be sent. If the characteristics of a predicting receiver are known, the transmitter can omit parts of the message in the knowledge that the receiver has the ability to recreate it. Thus all encoders must contain a model of the decoder.

The difference between the information rate and the overall bit rate is known as the redundancy. Compression systems are designed to eliminate as much of that redundancy as practicable or perhaps affordable. One way in which this can be done is to exploit statistical predictability in signals. The information content or entropy of a sample is a function of how different it is from the predicted value. Most signals have some degree of predictability. A sine wave is highly predictable, because all cycles look the same. According to Shannon's theory, any signal that is totally predictable carries no information. In the case of the sine wave this is clear because it represents a single frequency and so has no bandwidth.

At the opposite extreme a signal such as noise is completely unpredictable and as a result all codecs find noise difficult. There are two consequences of this characteristic. First, a codec designed using the statistics of real material should not be tested with random noise because it is not a representative test. Second, a codec that performs well with clean source material may perform badly with source material containing superimposed noise. Most practical compression units require some form of pre-processing before the compression stage proper and appropriate noise reduction should be incorporated into the pre-processing if noisy signals are anticipated. It will also be necessary to restrict the degree of compression applied to noisy signals.

All real signals fall mid-way between the extremes of total predictability and total unpredictability or noisiness. If the bandwidth (set by the sampling rate) and the dynamic range (set by the word length) of the transmission system are used to delineate an area, this sets a limit on the information capacity of the system. Figure 2.36(a) shows that most real signals only occupy part of that area. The signal may not contain all frequencies, or it may not have full dynamics at certain frequencies.

Entropy can be thought of as a measure of the actual area occupied by the signal. This is the area that must be transmitted if there are to be no subjective differences or artefacts in the received signal. The remaining area is called the redundancy because it adds nothing to the information conveyed. Thus an ideal coder could be imagined which miraculously sorts out the entropy from the redundancy and only sends the former. An ideal decoder would then recreate the original impression of the information quite perfectly.

As the ideal is approached, the coder complexity and the latency or delay both rise. Figure 2.36(b) shows how complexity increases with compression factor and (c) shows how increasing the codec latency can

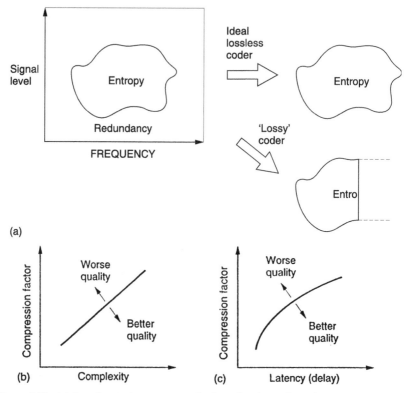

Figure 2.36 (a) A perfect coder removes only the redundancy from the input signal and results in subjectively lossless coding. If the remaining entropy is beyond the capacity of the channel some of it must be lost and the codec will then be lossy. An imperfect coder will also be lossy as it falls to keep all entropy. (b) As the compression factor rises, the complexity must also rise to maintain quality. (c) High compression factors also tend to increase latency or delay through the system.

improve the compression factor. Obviously we would have to provide a channel that could accept whatever entropy the coder extracts in order to have transparent quality. As a result, moderate coding gain that only removes redundancy need not cause artefacts and results in systems described as subjectively lossless.

If the channel capacity is not sufficient for that, then the coder will have to discard some of the entropy and with it useful information. Larger coding gain that removes some of the entropy must result in artefacts. It will also be seen from Figure 2.36 that an imperfect coder will fail to separate the redundancy and may discard entropy instead, resulting in artefacts at a suboptimal compression factor.

A single variable-rate transmission is unrealistic in broadcasting where fixed channel allocations exist. The variable-rate requirement can be met by combining several compressed channels into one constant rate transmission in a way that flexibly allocates data rate between the channels. Provided the material is unrelated, the probability of all channels reaching peak entropy at once is very small and so those channels that are at one instant passing easy material will free up transmission capacity for those channels that are handling difficult material. This is the principle of statistical multiplexing.

Lossless codes are less common for audio and video coding where perceptive codes are permissible. The perceptive codes often obtain a coding gain by shortening the word length of the data representing the signal waveform. This must increase the noise level and the trick is to ensure that the resultant noise is placed at frequencies where human senses are least able to perceive it. As a result although the received signal is measurably different from the source data, it can appear the same to the human listener or viewer at moderate compression factors. As these codes rely on the characteristics of human sight and hearing, they can only fully be tested subjectively.

The compression factor of such codes can be set at will by choosing the bit rate of the compressed data. Whilst mild compression will be undetectable, with greater compression factors, artefacts become noticeable. Figure 2.36 shows this to be inevitable from entropy considerations.

2.13   Introduction to audio compression

The human auditory system (HAS) is complex and at various times works in the frequency domain to analyse timbre or in the time domain to localize sound sources. In practice the precision of the HAS is finite in both domains and if this precision is known, it is possible to render signals less precise than their original condition, but which still seem as precise to the ear. In practice few audio compression systems are used in his way. Instead lower bit rates are used and the coder attempts to minimize the audible damage.

There are a number of coding tools available, but none of these is appropriate for all circumstances. Consequently practical coders will use combinations of different tools, usually selecting the most appropriate for the type of input.

The simplest coding tool is companding: a digital parallel of the noise reducers used in analog tape recording. Figure 2.37(a) shows that in companding systems the input signal level is monitored. Whenever the input level falls below maximum, it is amplified at the coder. The gain that was applied at the coder is added to the data stream so that the decoder can apply an equal attenuation. The advantage of companding is that the signal is kept as far away from the noise floor as possible. In analog noise reduction this is used to maximize the SNR of a tape recorder, whereas in

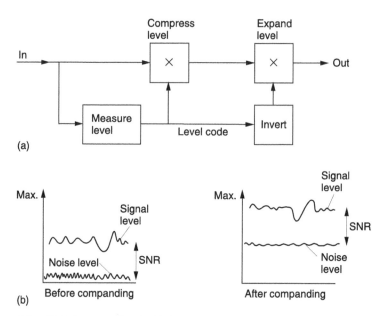

Figure 2.37 Digital companding. In (a) the encoder amplifies the input to maximum level and the decoder attenuates by the same amount. (b) In a companded system, the signal is kept as far as possible above the noise caused by shortening the sample word length.

digital compression it is used to keep the signal level as far as possible above the distortion introduced by various coding steps.

One common way of obtaining coding gain is to shorten the word length of samples so that fewer bits need to be transmitted. Figure 2.37(b) shows that when this is done, the distortion will rise by 6 dB for every bit removed. This is because removing a bit halves the number of quantizing intervals which then must be twice as large, doubling the error amplitude.

Clearly if this step follows the compander of (a), the audibility of the distortion will be minimized. As an alternative to shortening the word length, the uniform quantized PCM signal can be converted to a non-uniform format. In non-uniform coding, shown at (c), the size of the quantizing step rises with the magnitude of the sample so that the distortion level is greater when higher levels exist.

In sub-band coding, the audio spectrum is split into many different frequency bands. Once this has been done, each band can be individually processed. In real audio signals many bands will contain lower-level signals than the loudest one. Individual companding of each band will be more effective than broadband companding. Sub-band coding also allows the level of distortion products to be raised selectively so that distortion is created only at frequencies where spectral masking will be effective.

Transform coding is an extreme case of sub-band coding in which the sub-bands have become so narrow that they can be described by one coefficient.

Prediction is a coding tool in which the coder attempts to predict or anticipate the value of a future parameter from those already known to both encoder and decoder. It can be used in the time domain or the frequency domain. When used in the time domain, a predictor attempts to predict the value of the next audio sample. When used in the frequency domain, the predictor attempts to predict the value of the next frequency coefficient from those already sent.

The prediction will be subtracted from the actual value to obtain the prediction error, also known as a residual, which is transmitted. The decoder also contains a predictor that runs from the same signal history as the predictor in the encoder and will thus make the same prediction. By adding the residual, the decoder's predictor will obtain the correct sample value. Provided the prediction error is transmitted intact, it will be clear that predictive coding is lossless.

In MPEG Layer 1 audio coding, the input is split into 32 bands and individually companded in each. A Layer 1 block is thus quite simple, containing 32 gain factors and 32 word length codes to allow deserialization of the 32 sets of variable length samples.

In MPEG Layer 2 coding, the same number of sub-bands is used, but further coding gain is possible because redundancy in the gain factors is explored allowing the same parameter to be used for several blocks.

In MPEG Layer 3 coding, a transform is performed that decomposes the spectrum into 192 or 576 coefficients, depending on the transient content of the signal. These coefficients are then subject to non-uniform quantizing followed by a lossless mathematical packing algorithm.

The MPEG AAC (advanced audio coding) algorithm represents an improvement over the earlier codecs as it uses prediction that can adaptively switch between time and frequency domain to give better quality for a given bit rate.

2.14 Introduction to video compression

Video signals exist in four dimensions: these are the attributes of the sample, the horizontal and vertical spatial axes and the time axis. Compression can be applied in any or all of those four dimensions. MPEG-2 assumes eight-bit colour difference signals as the input, requiring rounding if the source is ten bit. The sampling rate of the colour signals is less than that of the luminance. This is done by a down-sampling of the colour samples horizontally and generally vertically as well. Essentially an MPEG-2 system has three parallel simultaneous channels, one for luminance and two for colour difference, which after coding are multiplexed into a single bitstream.

Figure 2.38(a) shows that spatial redundancy is redundancy within a single image, for example repeated pixel values in a large area of blue sky. Temporal redundancy (b) exists between successive images.

Where temporal compression is used, the current picture is not sent in its entirety; instead the difference between the current picture and the previous picture is sent. The decoder already has the previous picture, and so it can add the difference to make the current picture. A difference picture is created by subtracting every pixel in one picture from the corresponding pixel in another pixel.

A difference picture is an image of a kind, although not a viewable one, and so should contain some kind of spatial redundancy. Figure 2.38(c) shows that MPEG-2 takes advantage of both forms of redundancy. Picture differences are spatially compressed prior to transmission. At the decoder the spatial compression is decoded to recreate the difference picture, then this difference picture is added to the previous picture to complete the decoding process.

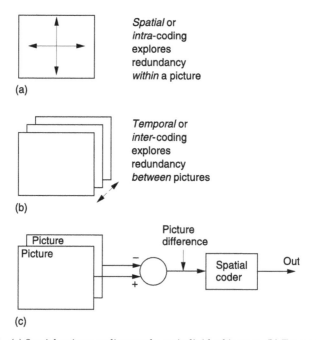

Figure 2.38 (a) Spatial or intra-coding works on individual images. (b) Temporal or inter-coding works on successive images. (c) In MPEG inter-coding is used to create difference images. These are the compressed spatially.

Whenever objects move they will be in a different place in successive pictures. This will result in large amounts of difference data. MPEG-2 overcomes the problem using motion compensation. The encoder contains a motion estimator that measures the direction and distance of motion between pictures and outputs these as vectors that are sent to the decoder. When the decoder receives the vectors it uses them to shift data in a previous picture to resemble the current picture more closely. Effectively the vectors are describing the optic flow axis of some moving screen area, along which axis the image is highly redundant. Vectors are bipolar codes that determine the amount of horizontal and vertical shift required.

In real images, moving objects do not necessarily maintain their appearance as they move. For example, objects may turn, move into shade or light, or move behind other objects. Consequently motion compensation can never be ideal and it is still necessary to send a picture difference to make up for any shortcomings in the motion compensation.

Figure 2.39 shows how this works. In addition to the motion-encoding system, the coder also contains a motion decoder. When the encoder outputs motion vectors, it also uses them locally in the same way that a real decoder will, and is able to produce a predicted picture based solely on the previous picture shifted by motion vectors. This is then subtracted from the actual current picture to produce a prediction error or residual which is an image of a kind that can be spatially compressed.

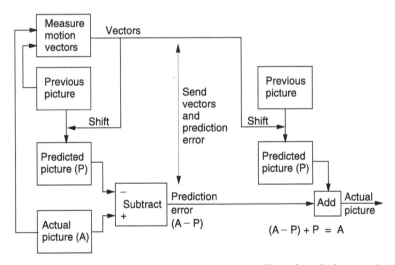

Figure 2.39 A motion-compensated compression system. The coder calculates motion vectors which are transmitted as well as being used locally to create a predicted picture. The difference between the predicted picture and the actual picture is transmitted as a prediction error.

The decoder takes the previous picture, shifts it with the vectors to recreate the predicted picture and then decodes and adds the prediction error to produce the actual picture. Picture data sent as vectors plus prediction error are said to be P coded.

The simple prediction system of Figure 2.39 is of limited use as in the case of a transmission error, every subsequent picture would be affected. Channel switching in a television set would also be impossible. In practical systems a modification is required. The approach used in MPEG is that periodically some absolute picture data is transmitted in place of difference data.

This absolute picture data, known as *I* or intra pictures, is interleaved with pictures which are created using difference data, known as *P* or predicted pictures. The *I* pictures require a large amount of data, whereas the *P* pictures require fewer data. As a result the instantaneous data rate varies dramatically and buffering has to be used to allow a constant transmission rate.

The *I* picture and all the *P* pictures prior to the next *I* picture are called a group of pictures (GOP). For a high compression factor, a large number of *P* pictures should be present between *I* pictures, making a long GOP. However, a long GOP delays recovery from a transmission error. The compressed bitstream can only be edited at *I* pictures.

Bi-directional coding is shown in Figure 2.40. Where moving objects reveal a background this is completely unknown in previous pictures and forward prediction fails. However, more of the background is visible in later pictures. In the centre of the diagram, a moving object has revealed

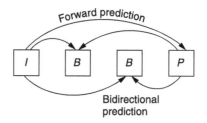

I = Intra- or spatially coded 'anchor' picture

P = Forward predicted. Coder sends difference between *I* and *P* decoder. Adds difference to create *P*

B = Bidirectionally coded picture can be coded from a previous *I* or *P* picture or a later *I* or *P* picture. *B* pictures are not coded from each other

Figure 2.40 In bi-directional coding, a number of *B* pictures can be inserted between periodic forward predicted pictures. See text.

some background. The previous picture can contribute nothing, whereas the next picture contains all that is required.

Bi-directional coding uses a combination of motion compensation and the addition of a prediction error. This can be done by forward prediction from a previous picture or backward prediction from a subsequent picture. It is also possible to use an average of both forward and backward prediction. On noisy material this may result in some reduction in bit rate. The technique is also a useful way of portraying a dissolve.

Typically two *B* pictures are inserted between *P* pictures or between *I* and *P* pictures. As can be seen, *B* pictures are never predicted from one another, only from *I* or *P* pictures. A typical GOP for broadcasting purposes might have the structure *IBBPBBPBBPBB*. Note that the last *B* pictures in the GOP require the *I* picture in the next GOP for decoding and so the GOPs are not truly independent. Independence can be obtained by creating a closed GOP which may contain *B* pictures but which ends with a *P* picture.

Bi-directional coding is very powerful. Figure 2.41 is a constant quality curve showing how the bit rate changes with the type of coding. On the left, only *I* or spatial coding is used, whereas on the right an *IBBP* structure is used having two bi-directionally coded pictures in between a spatially coded picture (*I*) and a forward predicted picture (*P*). Note how for the same quality the system that only uses spatial coding needs two and a half times the bit rate that the bi-directionally coded system needs.

Clearly information in the future has yet to be transmitted and so is not normally available to the decoder. MPEG-2 gets around the problem by sending pictures in the wrong order. Picture reordering requires delay in

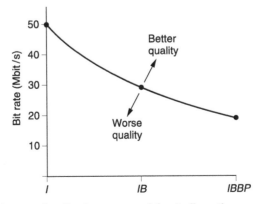

Figure 2.41 Bi-directional coding is very powerful as it allows the same quality with only 40% of the bit rate of intra-coding. However, the encoding and decoding delays must increase. Coding over a longer time span is more efficient but editing is more difficult.

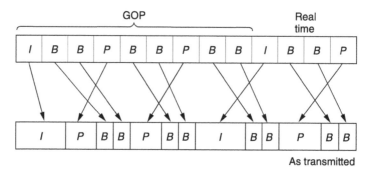

Figure 2.42 Comparison of pictures before and after compression showing sequence change and varying amount of data needed by each picture type. *I, P, B* pictures use unequal amounts of data.

the encoder and a delay in the decoder to put the order right again. Thus the overall codec delay must rise when bi-directional coding is used. This is quite consistent with Figure 2.36 in which it was shown that as the compression factor rises the latency must also rise.

Figure 2.42 shows that although the original picture sequence is *IBBPBBPBBIBB* ..., this is transmitted as *IPBBPBBIBB* ... so that the future picture is already in the decoder before bi-directional decoding begins. Note that the *I* picture of the next GOP is actually sent before the last *B* pictures of the current GOP.

Figure 2.42 also shows that the amount of data required by each picture is dramatically different. *I* pictures have only spatial redundancy and so need a lot of data to describe them. *P* pictures need fewer data because they are created by shifts of *I* picture using vectors and then adding a prediction error picture. *B* pictures need the least data of all because they can be created from *I* or *P*.

With pictures requiring a variable length of time to transmit, arriving in the wrong order, the decoder needs some help. This takes the form of picture-type flags and time stamps.

References

1. Watkinson, J.R., *Convergence in Broadcast and Communications Media*, Ch. 7, Focal Press, Oxford, (2001)
2. Betts, J.A., *Signal Processing Modulation and Noise*, Ch. 6, Hodder and Stoughton, Sevenoaks (1970)
3. Meyer, J., Time correction of anti-aliasing filters used in digital audio systems. *J. Audio Eng. Soc.*, vol. 32, pp. 132–137 (1984)
4. Blesser, B., Advanced A/D conversion and filtering: data conversion. In *Digital Audio*, ed. B. A. Blesser, B. Locanthi and T. G. Stockham Jr, pp. 37–53, Audio Engineering Society, New York (1983)

5. Lagadec, R., Weiss, D. and Greutmann, R., High-quality analog filters for digital audio. Presented at the *67th Audio Engineering Society Convention* (New York), preprint 1707 (B-4) (1980)

6. Hsu, S., The Kell factor: past and present. *SMPTE Journal*, vol. 95, pp. 206–214 (1986)

7. Jesty, L.C., The relationship between picture size, viewing distance and picture quality. *Proc. IEE*, vol. 105B, pp. 425–439 (1958)

8. Ishida, Y. *et al.*, A PCM digital audio processor for home use VTRs. Presented at *64th Audio Engineering Society Convention* (New York), preprint 1528 (1979)

9. Watkinson, J.R., *The Digital Video Tape Recorder*, Focal Press, Oxford (1994)

10. AES, AES recommended practice for professional digital audio applications employing pulse code modulation: preferred sampling frequencies. AES5–1984 (ANSI S4.28–1984), *J. Audio Eng. Soc.*, vol. 32, pp. 781–785 (1984)

11. Lipshitz, S.P. *et al.*, Quantization and dither: a theoretical survey. *J. Audio Eng. Soc.*, vol. 40, pp. 355–375 (1992)

12. Roberts, L.G., Picture coding using pseudo-random noise. *IRETrans. Inform. Theory*, vol. IT-8, pp. 145–154 (1962)

13. Watkinson, J.R., *The Art of Digital Audio*, Third Edn, Focal Press, Oxford (2001)

14. Watkinson, J.R., *The Art of Digital Video*, Third Edn, Focal Press, Oxford (2000)

15. Vanderkooy, J. and Lipshitz, S.P., Digital dither. Presented at the *81st Audio Engineering Society Convention* (Los Angeles), preprint 2412(C-8) (1986)

16. Watkinson, J.R., *The MPEG Handbook*, Focal Press, Oxford (2001)

3

Digital transmission

Digital transmission consists of converting data into a waveform suitable for the path along which they are to be sent allied to the adherence to some protocol or data structure that allows the receiving device correctly to interpret the data.

In this chapter the fundamentals of digital transmission are introduced along with descriptions of the coding and error-correction techniques used in practical applications.

3.1 Introduction

The generic term for the path down which information is sent is the *channel*. In a transmission application, the channel may be a point-to-point cable, a network stage or a radio link.

In digital circuitry there is a great deal of noise immunity because the signal has only two states, which are widely separated compared with the amplitude of noise. In transmission this is not always the case. In real channels, the signal may *originate* with discrete states which change at discrete times, but the channel will treat it as an analog waveform and so it will not be *received* in the same form. Various frequency-dependent loss mechanisms will reduce the amplitude of the signal. Noise will be picked up as a result of stray electric fields or magnetic induction and the receiving circuitry will contribute some noise. As a result, the received voltage will have an infinitely varying state along with a degree of uncertainty due to the noise. Different frequencies can propagate at different speeds in the channel; this is the phenomenon of group delay. An alternative way of considering group delay is that there will be frequency-dependent phase shifts in the signal and these will result in uncertainty in the timing of pulses.

In digital circuitry, the signals are generally accompanied by a separate clock signal that reclocks the data to remove jitter as was shown in Chapter 1. In contrast, it is generally not feasible to provide a separate clock in transmission applications. A separate clock line would not only raise cost, but would be impractical because at high frequency it is virtually impossible to ensure that the clock cable propagates signals at the same speed as the data cable except over short distances. Such timing differences between parallel channels are known as skew.

The solution is to use a self-clocking waveform and the generation of this is a further essential function of the coding process. Clearly, if data bits are simply clocked serially from a shift register in so-called direct transmission this characteristic will not be obtained. If all the data bits are the same, for example all zeros, a common code in audio and video, there is no clock when they are serialized.

It is not the channel which is digital; instead the term describes the way in which the received signals are *interpreted*. When the receiver makes discrete decisions from the input waveform it attempts to reject the uncertainties in voltage and time. The technique of channel coding is one where transmitted waveforms are restricted to those that still allow the receiver to make discrete decisions despite the degradations caused by the analog nature of the channel.

3.2 Types of transmission channel

Transmission can be by electrical conductors, radio or optical fibre. Although these appear to be completely different, they are in fact just different examples of electromagnetic energy travelling from one place to another. If the energy is made time-variant, information can be carried.

Electromagnetic energy propagates in a manner that is a function of frequency, and our incomplete understanding requires it to be considered as electrons, waves or photons so that we can predict its behaviour in given circumstances.

At DC and at the low frequencies used for power distribution, electromagnetic energy is called electricity and needs to be transported completely inside conductors. It has to have a complete circuit to flow in, and the resistance to current flow is determined by the cross-sectional area of the conductor. The insulation around the conductor and the spacing between the conductors has no effect on the ability of the conductor to pass current. At DC an inductor appears to be a short circuit, and a capacitor appears to be an open circuit.

As frequency rises, resistance is exchanged for impedance. Inductors display increasing impedance with frequency, capacitors show falling impedance. Electromagnetic energy increasingly tends to leave the conductor. The first symptom is the skin effect: the current flows only in the outside layer of the conductor effectively causing the resistance to rise.

As the energy is starting to leave the conductors, the characteristics of the space between them become important. This determines the impedance. A change of impedance causes reflections in the energy flow and some of it heads back towards the source. Constant impedance cables with fixed conductor spacing are necessary, and these must be suitably terminated to prevent reflections. The most important characteristic of the insulation is its thickness as this determines the spacing between the conductors.

As frequency rises still further, the energy travels less in the conductors and more in the insulation between them, and their composition becomes important and they begin to be called dielectrics. A poor dielectric like PVC absorbs high-frequency energy and attenuates the signal. So-called low-loss dielectrics such as PTFE are used, and one way of achieving low loss is to incorporate as much air into the dielectric as possible by making it in the form of foam or extruding it with voids.

High-frequency signals can also be propagated without a medium, and are called radio. As frequency rises further the electromagnetic energy is termed 'light' which can also travel without a medium, but can also be guided through a suitable medium. Figure 3.1(a) shows an early type of optical fibre in which total internal reflection is used to guide the light. It will be seen that the length of the optical path is a function of the angle at which the light is launched. Thus at the end of a long fibre sharp transitions would be smeared by this effect. Later optical fibres are made with a radius-dependent refractive index such that light diverging from the axis

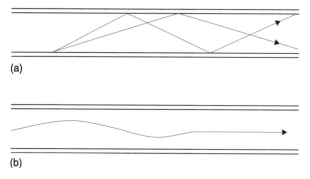

(a)

(b)

Figure 3.1   (a) Early optical fibres operated on internal reflection, and signals could take a variety of paths along the fibre, hence multi-mode. (b) Later fibres used graduated refractive index whereby light was guided to the centre of the fibre and only one mode was possible.

is automatically refracted back into the fibre. Figure 3.1(b) shows that in single-mode fibre light can only travel down one path and so the smearing of transitions is minimized.

3.3 Transmission lines

Frequency-dependent behaviour is the most important factor in deciding how best to harness electromagnetic energy flow for information transmission. It is obvious that the higher the frequency, the greater the possible information rate, but in general, losses increase with frequency, and flat frequency response is elusive. The best that can be managed is that over a narrow band of frequencies, the response can be made reasonably constant with the help of equalization. Unfortunately raw data when serialized have an unconstrained spectrum. Runs of identical bits can produce frequencies much lower than the bit rate would suggest. One of the essential steps in a transmission system is to modify the spectrum of the data into something more suitable.

At moderate bit rates, say a few megabits per second, and with moderate cable lengths, say a few metres, the dominant effect will be the capacitance of the cable due to the geometry of the space between the conductors and the dielectric between. The capacitance behaves under these conditions as if it were a single capacitor connected across the signal. The effect of the series source resistance and the parallel capacitance is that signal edges or transitions are turned into exponential curves. This happens because the capacitance is effectively being charged and discharged through the source impedance.

As cable length increases, the capacitance can no longer be lumped as if it were a single unit; it has to be regarded as being distributed along the cable. With rising frequency, the cable inductance also becomes significant, and it too is distributed.

The cable is now a transmission line and pulses travel down it as current loops that roll along as shown in Figure 3.2. If the pulse is positive, as it is launched along the line, it will charge the dielectric locally as at (a). As the pulse moves along, it will continue to charge the local dielectric as at (b). When the driver finishes the pulse, the trailing edge of the pulse follows the leading edge along the line. The voltage of the dielectric charged by the leading edge of the pulse is now higher than the voltage on the line, and so the dielectric discharges into the line as at (c). The current flows forward as it is in fact the same current that is flowing into the dielectric at the leading edge. There is thus a loop of current rolling down the line flowing forward in the 'hot' wire and backwards in the return.

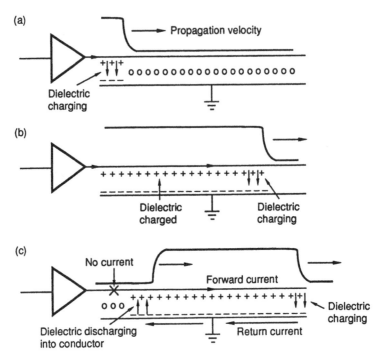

Figure 3.2   A transmission line conveys energy packets which appear with respect to the dielectric. In (a) the driver launches a pulse which charges the dielectric at the beginning of the line. As it propagates the dielectric is charged further along as in (b). When the driver ends the pulse, the charged dielectric discharges into the line. A current loop is formed where the current in the return loop flows in the opposite direction to the current in the 'hot' wire.

The constant to-ing and fro-ing of charge in the dielectric results in dielectric loss of signal energy. Dielectric loss increases with frequency and so a long transmission line acts as a filter. Thus the term 'low-loss' cable refers primarily to the kind of dielectric used. For serial digital high definition the dielectric loss is an important factor in the length of cable that can be used.

Transmission lines that transport energy in this way have a characteristic impedance due to interplay of the inductance along the conductors with the parallel capacitance. One consequence of that transmission mode is that correct termination or matching is required between the line and both the driver and the receiver. When a line is correctly matched, the rolling energy rolls straight out of the line into the load and the maximum energy is available. If the impedance presented by the load is incorrect, there will be reflections from the mismatch. An open circuit will reflect all the energy back in the same polarity as the original, whereas a short circuit will reflect all the energy back in the opposite polarity. Thus impedances above or below the correct value will have a tendency towards reflections whose magnitude depends upon the degree of mismatch and whose polarity depends upon whether the load is too high or too low. In practice it is the need to avoid reflections that is the most important reason to terminate correctly. Devices are specified by the return loss they exhibit and each interface standard will contain a return loss limit.

A perfectly square pulse contains an indefinite series of harmonics, but the higher ones suffer progressively more loss. A square pulse at the driver becomes less and less square with distance as Figure 3.3 shows. The harmonics are progressively lost until in the extreme case all that is

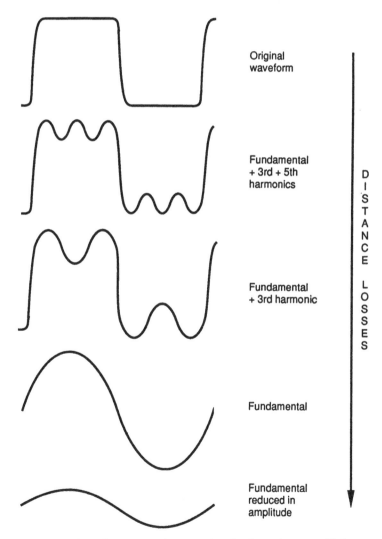

Original waveform

Fundamental + 3rd + 5th harmonics

Fundamental + 3rd harmonic

Fundamental

Fundamental reduced in amplitude

DISTANCE

LOSSES

Figure 3.3 A signal may be square at the transmitter, but losses increase with frequency, and as the signal propagates, more of the harmonics are lost until only the fundamental remains. The amplitude of the fundamental then falls with further distance.

left is the fundamental. A transmitted square wave is received as a sine wave. Fortunately data can still be recovered from the fundamental signal component.

Once all the harmonics have been lost, further losses cause the amplitude of the fundamental to fall. The effect worsens with distance and it is necessary to ensure that data recovery is still possible from a signal of unpredictable level.

3.4 Equalization and data separation

The characteristics of most channels are that signal loss occurs which increases with frequency. This has the effect of slowing down rise times and thereby sloping off edges. If a signal with sloping edges is sliced, the time at which the waveform crosses the slicing level will be changed, and this causes jitter. Figure 3.4 shows slicing a sloping waveform in the presence of baseline wander causes more jitter.

On a long cable, high-frequency roll-off can cause sufficient jitter to move a transition into an adjacent bit period. This is called intersymbol interference and the effect becomes worse in signals having greater asymmetry, i.e. short pulses alternating with long ones. The effect can be reduced by the application of equalization, which is typically a high-frequency boost, and by choosing a channel code which has restricted asymmetry.

The important step of information recovery at the receiver is known as data separation. The data separator is rather like an analog-to-digital convertor because the two processes of sampling and quantizing are both present. In the time domain, the sampling clock is derived from the clock content of the channel waveform. In the voltage domain, the process of *slicing* converts the analog waveform from the channel back into a binary representation. The slicer is thus a form of quantizer that typically only has one-bit resolution, although multi-level signalling will also be found. The slicing process makes a discrete decision about the voltage of the incoming signal in order to reject noise. The sampler makes discrete

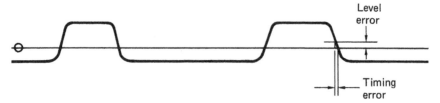

Figure 3.4   A DC offset can cause timing errors.

decisions along the time axis in order to reject jitter. These two processes will be described in detail.

3.5 Slicing and jitter rejection

The slicer is implemented with a comparator that has analog inputs but a binary output. In a cable receiver, this may follow the equalizer. The signal voltage is compared with the midway voltage, known as the threshold, baseline or slicing level by the comparator. If the signal voltage is above the threshold, the comparator outputs a high level, if below, a low level results. Figure 3.5 shows some waveforms associated with a slicer.

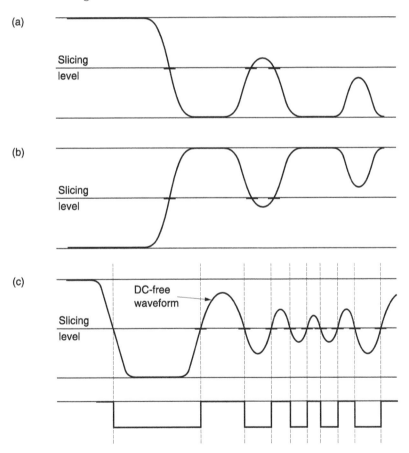

Figure 3.5 Slicing a signal which has suffered losses works well if the duty cycle is even. If the duty cycle is uneven, as at (a), timing errors will become worse until slicing fails. With the opposite duty cycle, the slicing fails in the opposite direction as at (b). If, however, the signal is DC free, correct slicing can continue even in the presence of serious losses, as (c) shows.

At (a) the transmitted waveform has an uneven duty cycle. The DC component, or average level, of the signal is received with high amplitude, but the pulse amplitude falls as the pulse gets shorter. Eventually the waveform cannot be sliced.

At (b) the opposite duty cycle is shown. The signal level drifts to the opposite polarity and once more slicing is impossible. The phenomenon is called baseline wander and will be observed with any signal whose average voltage is not the same as the slicing level.

At (c) it will be seen that if the transmitted waveform has a relatively constant average voltage, slicing remains possible up to high frequencies even in the presence of serious amplitude loss, because the received waveform remains symmetrical about the baseline.

It is clearly not possible simply to serialize data in a shift register for so-called direct transmission, because successful slicing can only be obtained if the number of ones is equal to the number of zeros; there is little chance of this happening consistently with real data. Instead, a modulation code or channel code is necessary. This converts the data into a waveform that is DC-free or nearly so for the purpose of transmission.

The slicing threshold level is naturally zero in a bipolar system such as a cable. When the amplitude falls it does so symmetrically and slicing continues. The same is not true of optical fibres in which the receiver responds to intensity and therefore produces a unipolar output. If the received waveform is sliced directly, the threshold cannot be zero, but must be some level approximately half the amplitude of the signal as shown in Figure 3.6(a).

Unfortunately when the signal level falls due to losses, it falls towards zero and not towards the slicing level. The threshold will no longer be appropriate for the signal as can be seen at (b). This can be overcome by using a DC-free coded waveform. If a series capacitor is connected to the unipolar signal from an optical receiver, the waveform is rendered bipolar because the capacitor blocks any DC component in the signal. The DC-free channel waveform passes through unaltered. If an amplitude loss is suffered, (c) shows that the resultant bipolar signal now reduces in amplitude about the slicing level and slicing can continue.

The binary waveform at the output of the slicer will be a replica of the transmitted waveform, except for the addition of jitter or time uncertainty in the position of the edges due to noise, baseline wander, intersymbol interference and imperfect equalization.

Binary circuits reject noise by using discrete voltage levels which are spaced further apart than the uncertainty due to noise. In a similar manner, digital coding combats time uncertainty by making the time axis discrete using events, known as transitions, spaced apart at integer

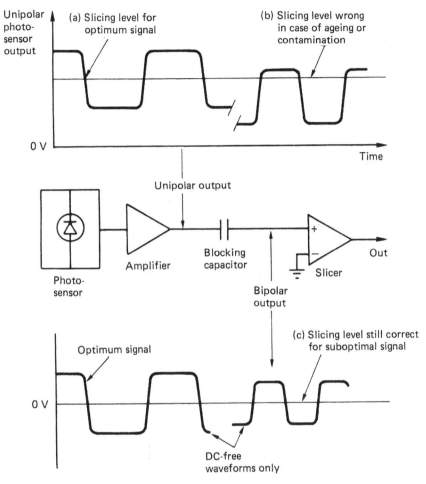

Figure 3.6 (a) Slicing a unipolar signal requires a non-zero threshold. (b) If the signal amplitude changes, the threshold will then be incorrect. (c) If a DC-free code is used, a unipolar waveform can be converted to a bipolar waveform using a series capacitor. A zero threshold can be used and slicing continues with amplitude variations.

multiples of some basic time period, called a detent, which is larger than the typical time uncertainty. Figure 3.7 shows how this jitter-rejection mechanism works. All that matters is to identify the detent in which the transition occurred. Exactly where it occurred within the detent is of no consequence.

As ideal transitions occur at multiples of a basic period, an oscilloscope, which is repeatedly triggered on a channel-coded signal carrying random data, will show an eye pattern if connected to the output of the equalizer. Study of the eye pattern reveals how well the coding used suits the channel. In the case of transmission, with a short cable, the losses will be small, and

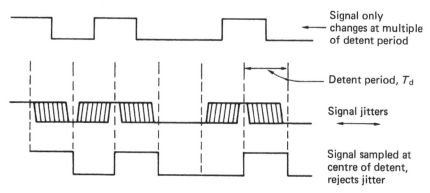

Figure 3.7 A certain amount of jitter can be rejected by changing the signal at multiples of the basic detent period T_d.

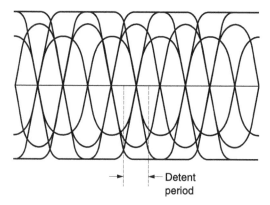

Figure 3.8 A transmitted waveform will appear like this on an oscilloscope as successive parts of the waveform are superimposed on the tube. When the waveform is rounded off by losses, diamond-shaped eyes are left in the centre, spaced apart by the detent period.

the eye opening will be virtually square except for some edge-sloping due to cable capacitance. As cable length increases, the harmonics are lost and the remaining fundamental gives the eyes a diamond shape.

Noise closes the eyes in a vertical direction, and jitter closes the eyes in a horizontal direction, as in Figure 3.8. If the eyes remain sensibly open, data separation will be possible. Clearly, more jitter can be tolerated if there is less noise, and vice versa. If the equalizer is adjustable, the optimum setting will be where the greatest eye opening is obtained.

In the centre of the eyes, the receiver must make binary decisions at the channel bit rate about the state of the signal, high or low, using the slicer output. As stated, the receiver is sampling the output of the slicer, and it needs to have a sampling clock in order to do that. In order to give the best rejection of noise and jitter, the clock edges that operate the sampler must be in the centre of the eyes.

As has been stated, a separate clock is not practicable in recording or transmission. A fixed-frequency clock at the receiver is of no use. Even if it were sufficiently stable, it would not know the phase at which to run.

The only way in which the sampling clock can be obtained is to use a phase-locked loop to regenerate it from the clock content of the self-clocking channel-coded waveform. In phase-locked loops, the voltage-controlled oscillator is driven by a phase error measured between the output and some reference, such that the output eventually has the same frequency as the reference. If a divider is placed between the VCO and the phase comparator, the VCO frequency can be made to be a multiple of the reference. This also has the effect of making the loop more heavily damped. If a channel-coded waveform is used as a reference to a PLL, the loop will be able to make a phase comparison whenever a transition arrives and will run at the channel bit rate. When there are several detents between transitions, the loop will *flywheel* at the last known frequency and phase until it can rephase at a subsequent transition. Thus a continuous clock is re-created from the clock content of the channel waveform. In a recorder, if the speed of the medium should change, the PLL will change frequency to follow. Once the loop is locked, clock edges will be phased with the average phase of the jittering edges of the input waveform. If, for example, rising edges of the clock are phased to input transitions, then falling edges will be in the centre of the eyes. If these edges are used to clock the sampling process, the maximum jitter and noise can be rejected. The output of the slicer when sampled by the PLL edge at the centre of an eye is the value of a channel bit. Figure 3.9 shows the complete clocking system of a channel code from encoder to data separator.

Clearly, data cannot be separated if the PLL is not locked, but it cannot be locked until it has seen transitions for a reasonable period. In interfaces, transmission can be continuous and there is no difficulty remaining in lock indefinitely. There will simply be a short delay on first applying the signal before the receiver locks to it.

One potential problem area that is frequently overlooked is to ensure that the VCO in the receiving PLL is correctly centred. If it is not, it will be running with a static phase error and will not sample the received waveform at the centre of the eyes. The sampled bits will be more prone to noise and jitter errors. VCO centring is considered in Chapter 8. Many interface receivers have such an adjustment, although recent receiving chip sets may incorporate self-adjustment.

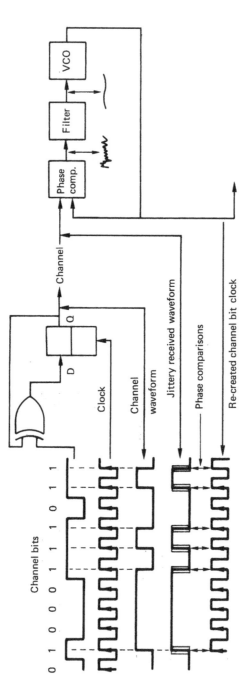

Figure 3.9 The complete clock path of a channel coding system.

3.6 Channel coding

It is not practicable simply to serialize raw data in a shift register for transmission, except over relatively short distances. Practical systems require the use of a modulation scheme, known as a channel code, which expresses the data as waveforms which are self-clocking in order to reject jitter, to separate the received bits and to avoid skew on separate clock lines. The coded waveforms should ideally be DC-free or nearly so to enable slicing in the presence of losses and have a narrower spectrum than the raw data both for economy and to make equalization easier.

Jitter causes uncertainty about the time at which a particular event occurred. The frequency response of the channel then places an overall limit on the spacing of events in the channel. Particular emphasis must be placed on the interplay of bandwidth, jitter and noise, which will be shown here to be the key to the design of a successful channel code.

Figure 3.10 shows that a channel coder is necessary at the transmitter, and that a decoder, known as a data separator, is necessary at the receiver. The output of the channel coder is generally a logic-level signal that contains a 'high' state when a transition is to be generated. The waveform generator produces the transitions in a signal whose level and impedance are suitable for driving the channel. The signal may be bipolar or unipolar as appropriate.

One of the fundamental parameters of a transmission code is the efficiency. This can be thought of as the ratio between the Nyquist rate of the data (one-half the bit rate) and the channel bandwidth needed. Efficiency is measured in bits/sec/Hz. Another way of considering the efficiency is that it is the worst-case ratio of the number of data bits transmitted to the number of transitions in the channel.

Some codes eliminate DC entirely. Some codes can reduce the channel bandwidth by lowering the upper spectral limit. This permits greater efficiency (measured in bits/sec/Hz), usually at the expense of noise and jitter rejection. Other codes narrow the spectrum, by raising the lower limit. A code with a narrow spectrum has a number of advantages. The reduction in asymmetry will reduce peak shift and data separators can lock more readily because the range of frequencies in the code is smaller. In theory the narrower the spectrum, the less noise will be suffered, but this is only achieved if filtering is employed. Filters can easily cause phase errors that will nullify any gain.

A convenient definition of a channel code (for there are certainly others) is 'A method of modulating real data such that they can be reliably received despite the shortcomings of a real channel, whilst making maximum economic use of the channel capacity.' The basic time periods of a channel-coded waveform are called positions or detents, in which the

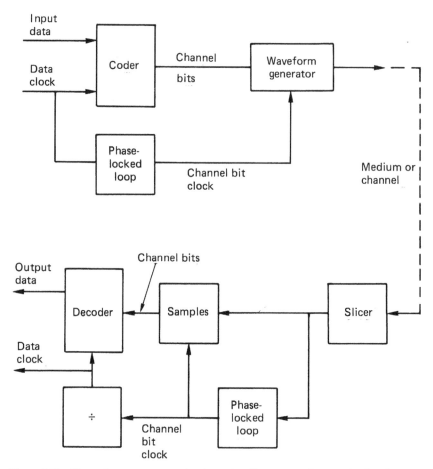

Figure 3.10 The major components of a channel coding system. See text for details.

transmitted voltage will be reversed or stay the same. The symbol used for the units of channel time is T_d.

As jitter is such an important issue in digital transmission, a parameter has been introduced to quantify the ability of a channel code to reject time instability. This parameter, the jitter margin, also known as the window margin or phase margin (T_w), is defined as the permitted range of time over which a transition can still be received correctly, divided by the data bit-cell period (T).

Since equalization is often difficult in practice, a code having a large jitter margin will sometimes be used because it resists the effects of inter-symbol interference well. Such a code may achieve a better performance in practice than a code with a higher efficiency but poor jitter performance.

A more realistic comparison of code performance will be obtained by taking into account both efficiency and jitter margin.

3.7 Simple codes

The FM code, also known as Manchester code or bi-phase mark code, shown in Figure 3.11(a) was the first practical self-clocking binary code and it is suitable for both transmission and recording. It is DC-free and very easy to encode and decode. It is the code specified for the AES/EBU digital audio interconnect standard.

In FM there is always a transition at the bit-cell boundary acting as a clock. For a data one, there is an additional transition at the bit-cell centre. Figure 3.11(a) shows each data bit to be represented by two channel bits. For a data zero, they will be 10, and for a data one they will be 11. Since the first bit is always one, it conveys no information, and is responsible for the efficiency of only one-half. Since there can be two transitions for each data bit, the jitter margin can only be half a bit. The high clock content of FM does, however, mean that data recovery is possible over a wide range of bit rates; hence the use in AES/EBU. The lowest frequency in FM is due to a stream of zeros and is equal to half the bit rate. The highest frequency is due to a stream of ones, and is equal to the bit rate. Thus the fundamentals of FM are within a band of one octave. Effective equalization is generally possible over such a band. FM is not polarity-conscious and can be inverted without changing the data.

Figure 3.11(b) shows how an FM coder works. Data words are loaded into the input shift register and clocked out at the data bit rate. Each data bit is converted to two channel bits in the codebook or look-up table. These channel bits are loaded into the output register. The output register is clocked twice as fast as the input register because there are twice as many channel bits as data bits. The ratio of the two clocks is called the code rate, in this case it is a rate one-half code. Ones in the serial channel bit output represent transitions whereas zeros represent no change. The channel bits are fed to the waveform generator which is a one-bit delay, clocked at the channel bit rate, and an exclusive-OR gate. This changes state when a channel bit one is input. The result is a coded FM waveform where there is always a transition at the beginning of the data bit period, and a second optional transition whose presence indicates a one.

3.8 Group codes

Further improvements in coding rely on converting patterns of real data to patterns of channel bits with more desirable characteristics using a conversion table known as a codebook. If a data symbol of m bits is considered, it can have 2^m different combinations. As it is intended to discard

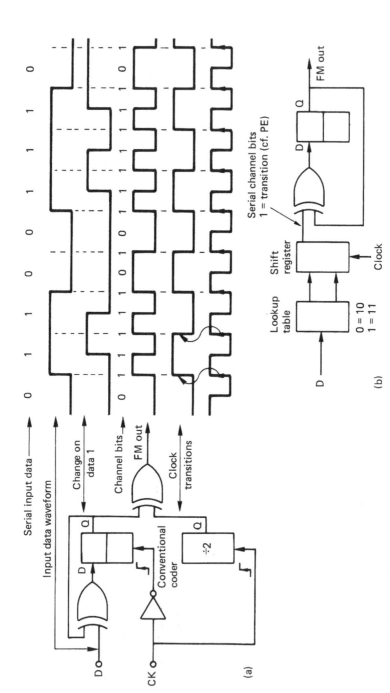

Figure 3.11 FM encoding.

_____

undesirable patterns to improve the code, it follows that the number of channel bits n must be greater than m. The number of patterns that can be discarded is:

$$2^n - 2^m$$

One name for the principle is group coding, and an important parameter is the code rate, defined as:

$$R = m/n$$

It will be evident that the jitter margin T_w is numerically equal to the code rate, and so a code rate near to unity is desirable. The patterns that are used in the codebook will be those giving the desired balance between clock content, bandwidth and DC content.

Figure 3.12 shows that the upper spectral limit can be made to be some fraction of the channel bit rate according to the minimum distance between ones in the channel bits. This is known as T_{min}, also referred to as the minimum transition parameter M and in both cases is measured in data bits T. It can be obtained by multiplying the number of channel detent periods between transitions by the code rate.

Figure 3.12 also shows the lower spectral limit to be influenced by the maximum distance between transitions T_{max}. This is also obtained by multiplying the maximum number of detent periods between transitions by the code rate. The length of time between channel transitions is known as the *run length*. Another name for this class is the run-length-limited (RLL) codes[1].

The DVB-ASI interface uses the electrical interface of SDI, but the statistics of MPEG compressed data are not suitable for the standard SDI coding scheme. Instead DVB-ASI uses an 8–10 group code in which eight

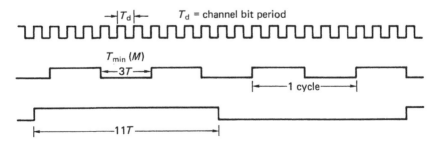

Figure 3.12 A channel code can control its spectrum by placing limits on T_{min} (M) and T_{max} which define upper and lower frequencies. The ratio of T_{max}/T_{min} determines the asymmetry of waveform and predicts DC content and peak shift.

data bits are converted to selected patterns of ten channel bits for transmission.

In the MADI (multi-channel audio interface) standard[2], a four-fifths rate code is used where groups of four data bits are represented by groups of five channel bits.

Four bits have 16 combinations whereas five bits have 32 combinations. Clearly only 16 out of these 32 are necessary to convey all the possible data. Figure 3.13 shows that the 16 channel bit patterns chosen are those having the smallest DC component combined with a high clock content. Adjacent ones are permitted in the channel bits, so there can be no violation of T_{min} at the boundary of two symbols. T_{max} is determined by the worst-case run of zeros at a symbol boundary and as $k = 3$, T_{max} is $16/5 = 3.2T$. The code is thus described as 0,3,4,5,1 and $L_c = 4T$.

The jitter resistance of a group code is equal to the code rate. For example, in 4/5 transitions cannot be closer than 0.8 of a data bit apart and so this represents the peak-to-peak jitter that can be rejected. The density ratio is also 0.8 so the FoM is 0.64; an improvement over FM.

A further advantage of group coding is that it is possible to have codes that have no data meaning. In MADI further channel bit patterns are used for packing and synchronizing. Packing, or stuffing, is the name given to dummy data sent when the real data rate is low in order to keep the channel frequencies constant. This is necessary so that fixed equalization can be used. The packing pattern does not decode to data and so it can easily be discarded at the receiver. Further details of MADI can be found in Chapter 4.

4 bit data	5 bit encoded data
0000	11110
0001	01001
0010	10100
0011	10101
0100	01010
0101	01011
0110	01110
0111	01111
1000	10010
1001	10011
1010	10110
1011	10111
1100	11010
1101	11011
1110	11100
1111	11101
SYNC	{ 11000 / 10001

Figure 3.13 The codebook of the 4/5 code of MADI. Note that a one represents a transition in the channel.

3.9   Randomizing and encryption

Randomizing is not a channel code, but a technique that can be used in conjunction with almost any channel code. It is widely used in digital audio and video broadcasting and in a number of transmission formats. The randomizing system is arranged outside the channel coder. Figure 3.14 shows that, at the encoder, a pseudo-random sequence is added modulo-2 to the serial data. This process makes the signal spectrum in the channel more uniform, drastically reduces T_{max} and reduces DC content. At the receiver the transitions are converted back to a serial bitstream to which the same pseudo-random sequence is again added modulo-2. As a result, the random signal cancels itself out to leave only the serial data, provided that the two pseudo-random sequences are synchronized to bit accuracy.

Many channel codes, especially group codes, display pattern sensitivity because some waveforms are more sensitive to peak shift distortion than others. Pattern sensitivity is only a problem if a sustained series of sensitive symbols needs to be recorded. Randomizing ensures that this cannot happen because it breaks up any regularity or repetition in the data. The data randomizing is performed by using the exclusive-OR function of the data and a pseudo-random sequence as the input to the channel coder. On replay the same sequence is generated, synchronized to bit accuracy, and the exclusive-OR of the replay bitstream and the sequence is the original data.

Clearly, the sync pattern cannot be randomized, since this causes a situation where it is not possible to synchronize the sequence for replay until the sync pattern is read, but it is not possible to read the sync pattern until the sequence is synchronized!

In networks, the randomizing may be block based, since this matches the block structure of the transmission protocol. Where there is no obvious block structure, convolutional or endless randomizing can be used. In convolutional randomizing, the signal sent down the channel is the serial

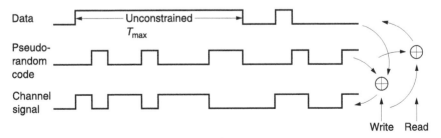

Figure 3.14   Modulo-2 addition with a pseudo-random code removes unconstrained runs in real data. Identical process must be provided on replay.

data waveform convolved with the impulse response of a digital filter. On reception the signal is deconvolved to restore the original data.

Convolutional randomizing is used in the serial digital interface (SDI) that carries 4:2:2 sampled video. Figure 3.15(a) shows that the filter is an infinite impulse response (IIR) filter having recursive paths from the output back to the input. As it is a one-bit filter its output cannot decay, and once excited, it runs indefinitely. Following the filter is a transition generator consisting of a one-bit delay and an exclusive-OR gate. An input 1 results in an output transition on the next clock edge. An input 0 results in no transition.

A result of the infinite impulse response of the filter is that frequent transitions are generated in the channel resulting in sufficient clock content for the phase-locked loop in the receiver.

Transitions are converted back to 1s by a differentiator in the receiver. This consists of a one-bit delay with an exclusive-OR gate comparing the input and the output. When a transition passes through the delay, the input and the output will be different and the gate outputs a 1 that enters the deconvolution circuit.

Figure 3.15(b) shows that in the deconvolution circuit a data bit is simply the exclusive-OR of a number of channel bits at a fixed spacing. The deconvolution is implemented with a shift register having the exclusive-OR gates connected in a reverse pattern to that in the encoder. The same effect as block randomizing is obtained, in that long runs are broken up and the DC content is reduced, but it has the advantage over block randomizing that no synchronizing is required to remove the randomizing, although it will still be necessary for deserialization. Clearly, the system will take a few clock periods to produce valid data after commencement of transmission, but this is no problem on a permanent wired connection where the transmission is continuous.

Where randomizing is used instead of a channel code, as in the serial digital interface, reliable operation depends on the statistics of the data. If the data to be transmitted have a certain combination of bits, the randomizing is defeated and the clock content is dramatically reduced. Such bit combinations are known as pathological sequences and may deliberately be generated for testing purposes. In video data, the probability of natural occurrence of pathological sequences is very small indeed, whereas in generic data it is not. This is why DVB-ASI cannot use this coding scheme.

In a randomized transmission, if the receiver is not able to re-create the pseudo-random sequence, the data cannot be decoded. This can be used as the basis for encryption in which only authorized users can decode transmitted data. In an encryption system, the goal is security whereas in a channel-coding system the goal is simplicity. Channel coders use pseudo-random sequences because these are economical to create using

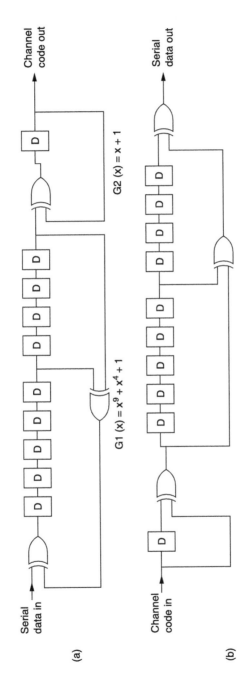

Figure 3.15 Convolutional randomizing encoder.

feedback shift registers. However, there are a limited number of pseudo-random sequences and it would be too easy to try them all until the correct one was found. Encryption systems use the same processes, but the key sequence that is added to the data at the encoder is truly random. This makes it much harder for unauthorized parties to access the data. Only a receiver in possession of the correct sequence can decode the channel signal. If the sequence is made long enough, the probability of stumbling across the sequence by trial and error can be made sufficiently small.

3.10 Synchronizing

Once the PLL in the data separator has locked to the clock content of the transmission, a serial channel bitstream and a channel bit clock will emerge from the sampler. In a group code, it is essential to know where a group of channel bits begins in order to assemble groups for decoding to data bit groups. In a randomizing system it is equally vital to know at what point in the serial data stream the words or samples commence. In serial transmission, channel bit groups or randomized data words are sent one after the other, one bit at a time, with no spaces in between, so that although the designer knows that a data packet contains, say, 128 bytes, the receiver simply finds 1024 bits in a row. If the exact position of the first bit is not known, then it is not possible to put all the bits in the right places in the right bytes; a process known as deserializing. The effect of sync slippage is devastating, because a one-bit disparity between the bit count and the bitstream will corrupt every symbol in the block.

The synchronization of the data separator and the synchronization to the packet format are two distinct problems, although they may be solved by the same sync pattern. Deserializing requires a shift register fed with serial data and read out once per word. The sync detector is simply a set of logic gates arranged to recognize a specific pattern in the register. The sync pattern is either identical for every packet or has a restricted number of versions and it will be recognized by the receiver circuitry and used to reset the bit count through the packet. Then by counting channel bits and dividing by the group size, groups can be deserialized and decoded to data groups. In a randomized system, the pseudo-random sequence generator is also reset. Then counting derandomized bits from the sync pattern and dividing by the word length enables the receiver to deserialize the data words.

Even if a specific code were excluded from the transmitted data so that it could be used for synchronizing, this cannot ensure that the same pattern cannot be falsely created at the junction between two allowable data words. Figure 3.16 shows how false synchronizing can occur due to concatenation.

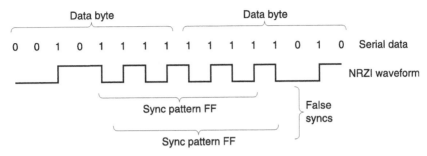

Figure 3.16 Concatenation of two words can result in the accidental generation of a word which is reserved for synchronizing.

It is thus not practical to search for a bit pattern that is a data code value. The problem is overcome in some synchronous systems by using the fact that sync patterns occur exactly once per packet and therefore contain redundancy. If the pattern is recognized at packet rate, a genuine sync condition exists. Sync patterns seen at other times must be false. Such systems take a few milliseconds before sync is achieved, but once achieved it should not be lost unless the transmission is seriously impaired.

In run-length-limited codes false syncs are not a problem. The sync pattern is no longer a data bit pattern but is a specific waveform. If the sync waveform contains run lengths that violate the normal coding limits, such run lengths cannot occur in encoded data, nor can they be interpreted as data. They can, however, be readily detected by the receiver.

In a group code there are many more combinations of channel bits than there are combinations of data bits. Thus after all data bit patterns have been allocated group patterns, there are still many unused group patterns which cannot occur in the data. With care, group patterns can be found which cannot occur due to the concatenation of any pair of groups representing data. These are then unique and can be used for synchronizing.

3.11 Basic error correction

There are many different types of transmission channel and consequently there will be many different mechanisms that may result in errors. Bit errors in audio cause unpleasant crackles. In video they cause 'sparkles' in the picture whose effect depends upon the significance of the affected bit. Errors in compressed data are more serious as they may cause the decoder to lose sync.

Errors may be caused by outside interference, or by Gaussian thermal noise in the channel and receiver. When group codes are used, a single defect in a group changes the group symbol and may cause errors up to the size of the group. Single-bit errors are therefore less common in group-coded channels. Inside equipment, data are conveyed on short wires and the noise environment is under the designer's control. With suitable design techniques, errors can be made effectively negligible whereas in communication systems, there is considerably less control of the electromagnetic environment. In networks, packets may be lost.

Irrespective of the cause, all these mechanisms cause one of two effects. There are large isolated corruptions, called error bursts, where numerous bits are corrupted all together in an otherwise error-free area, and there are random errors affecting single bits or symbols. Whatever the mechanism, the result will be that the received data will not be exactly the same as those sent. In binary the discrete bits will be each either right or wrong. If a binary digit is known to be wrong, it is only necessary to invert its state and then it must be right. Thus error correction itself is trivial; the hard part is working out *which* bits need correcting.

There are a number of terms having idiomatic meanings in error correction. The raw BER (bit error rate) is the error rate of the medium, whereas the residual or uncorrected BER is the rate at which the error-correction system fails to detect or miscorrects errors. In practical digital systems, the residual BER is negligibly small. If the error correction is turned off, the two figures become the same.

Error correction works by adding some bits to the data that are calculated from the data. This creates an entity called a codeword spanning a greater length of time than one bit alone. The statistics of noise means that whilst one bit may be lost in a codeword, the loss of the rest of the codeword because of noise is highly improbable. As will be described later in this chapter, codewords are designed to be able to correct totally a finite number of corrupted bits. The greater the time span over which the coding is performed, the greater will be the reliability achieved, although this does mean that an encoding delay will be experienced, along with a similar or greater decoding delay.

Shannon[3] disclosed that a message could be sent to any desired degree of accuracy provided that it is spread over a sufficient time span. Engineers have to compromise, because an infinite coding delay in the recovery of an error-free signal is not acceptable.

In some applications, such as AES/EBU and SDI, the requirements of production are such that no delay is acceptable. Such interfaces cannot use error correction and so are designed to be so robust that errors are negligibly infrequent in normal use. Instead of error correction, error detection is employed. If errors are detected, this implies that some maintenance is required.

If error correction is necessary as a practical matter, it is then only a small step to put it to maximum use. All error correction depends on adding bits to the original message, and this, of course, increases the number of bits to be sent, although it does not increase the information. It might be imagined that error correction is going to reduce efficiency, because bandwidth has to be found for all the extra bits. Nothing could be further from the truth. Once an error-correction system is used, the signal-to-noise ratio of the channel can be reduced, because the raised BER of the channel will be overcome by the error-correction system. Consequently the power of a digital transmitter can be reduced if error correction is used.

Figure 3.17 shows the broad subdivisions of error handling. The first stage might be called error avoidance and includes such measures as network rerouting to bypass faulty links. Properly terminating network cabling is also in this category. The data pass through the channel, which causes whatever corruptions it will. On receipt of the data the occurrence of errors is first detected, and this process must be extremely reliable, as it does not matter how effective the correction or how good the concealment algorithm, if it is not known that they are necessary. The detection of an error then results in a course of action being decided.

In the case of a file transfer, real-time operation is not required. A packet in error in a network may result in a retransmission. In many cases of digital video or audio replay a retransmission is not possible because the data are required in real time. In this case the solution is to encode the message using a system sufficiently powerful to correct the errors in real time. These are called forward error-correcting schemes (FEC). The term

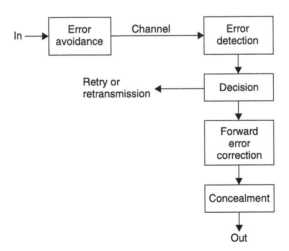

Figure 3.17 Error-handling strategies can be divided into avoiding errors, detecting errors and deciding what to do about them. Some possibilities are shown here. Of all these the detection is the most critical, as nothing can be done if the error is not detected.

'forward' implies that the transmitter does not need to take any action in the case of an error: the receiver will perform the correction.

3.12   Concealment by interpolation

There are some practical differences between data representing audio or video and generic data. Video and audio must be reproduced in real time, but have the advantage that there is a certain amount of redundancy in the information conveyed. Thus if an error cannot be corrected, it can be concealed. If a sample is lost, it is possible to obtain an approximation to it by interpolating between samples in the vicinity of the missing one. Clearly, concealment of any kind cannot be used with computer instructions or compressed data, although concealment can be applied after compressed signals have been decoded.

If there is too much corruption for concealment, the only course in video is repeat the previous field or frame in a freeze as it is unlikely that the corrupt picture is viewable. In audio the equivalent is muting.

In general, if use is to be made of concealment on receipt, the data must generally be reordered or shuffled prior to transmission. To take a simple example, odd-numbered samples are sent with a delay with respect to even-numbered samples. If a gross error occurs, depending on its timing, the result will be either corrupted odd samples or corrupted even samples, but it is most unlikely that both will be lost. Interpolation is then possible if the power of the correction system is exceeded. NICAM 728 uses such a system.

It should be stressed that corrected data are indistinguishable from the original and thus there can be no visible or audible artefacts. In contrast, concealment is only an approximation to the original information and could be detectable. In practical equipment, concealment occurs infrequently unless there is a defect requiring attention, and its presence is difficult to detect.

3.13   Parity

The error-detection and error-correction processes are closely related and will be dealt with together here. The actual correction of an error is simplified tremendously by the adoption of binary. As there are only two symbols, 0 and 1, it is enough to know that a symbol is wrong, and the correct value is obvious. Figure 3.18 shows a minimal circuit required for

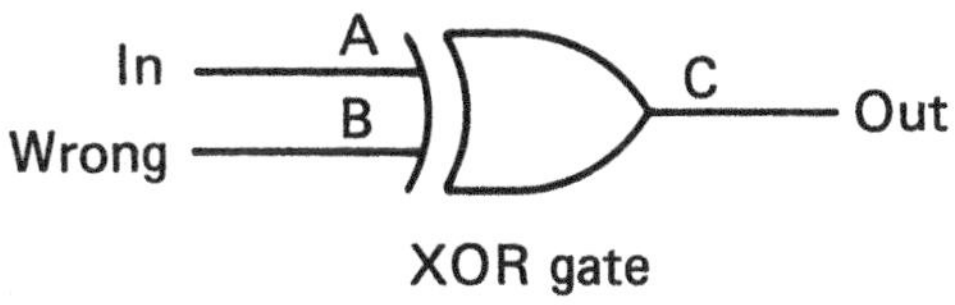

Truth table
of XOR gate

A	B	C
0	0	0
0	1	1
1	0	1
1	1	0

In ——— A
Wrong ——— B)) C ——— Out

XOR gate

$$A \oplus B = C$$

Figure 3.18 Once the position of the error is identified, the correction process in binary is easy.

correction once the bit in error has been identified. The XOR (exclusive-OR) gate shows up extensively in error correction and the figure also shows the truth table. One way of remembering the characteristics of this useful device is that there will be an output when the inputs are different. Inspection of the truth table will show that there is an even number of ones in each row (zero is an even number) and so the device could also be called an even parity gate. The XOR gate is also an adder in modulo-2.

Parity is a fundamental concept in error detection. In Figure 3.19, the example is given of a four-bit data word to be protected. If an extra bit is added to the word calculated in such a way that the total number of ones in the five-bit word is even, this property can be tested on receipt. The generation of the parity bit can be performed by a number of the ubiquitous XOR gates configured into what is known as a parity tree. In the figure, if a bit is corrupted, the received message will be seen no longer to have an even number of ones. If two bits are corrupted, the failure will be undetected. This example can be used to introduce much of the terminology of error correction. The extra bit added to the message carries no information of its own, since it is calculated from the other bits. It is therefore called a *redundant* bit.

The addition of the redundant bit gives the message a special property, i.e. the number of ones is even. A message having some special property *irrespective of the actual data content* is called a *codeword*. All error correction relies on adding redundancy to real data to form codewords for transmission. If any corruption occurs, the intention is that the received message will not have the special property; in other words, if the received message is not a codeword there has definitely been an error. The receiver can check for the special property without any prior knowledge of the data content. Thus the same check can be made on all received data. If the

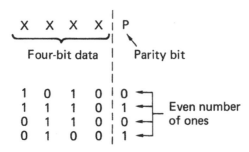

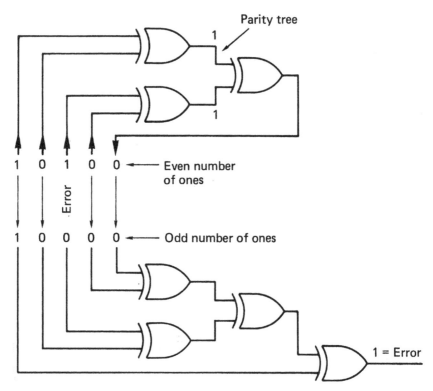

Figure 3.19   Parity checking adds up the number of ones in a word using, in this example, parity trees. One error bit and odd numbers of errors are detected. Even numbers of errors cannot be detected.

received message is a codeword, there probably has not been an error. The word 'probably' must be used because the figure shows that two bits in error will cause the received message to be a codeword, which cannot be discerned from an error-free message.

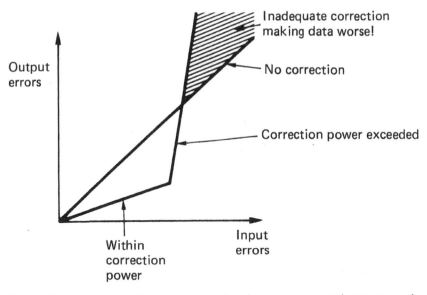

Figure 3.20 An error-correction system can only reduce errors at normal error rates at the expense of increasing errors at higher rates. It is most important to keep a system working to the left of the knee in the graph.

If it is known that generally the only failure mechanism in the channel in question is loss of a single bit, it is *assumed* that receipt of a codeword means that there has been no error. If there is a probability of two error bits, which becomes very nearly the probability of failing to detect an error, since all odd numbers of errors will be detected, and a four-bit error is much less likely. It is paramount in all error-correction systems that the protection used should be appropriate for the probability of errors to be encountered. An inadequate error-correction system is actually worse than not having any correction. Error correction works by trading probabilities. Error-free performance with a certain error rate is achieved at the expense of performance at higher error rates. Figure 3.20 shows the effect of an error-correction system on the residual BER for a given raw BER. It will be seen that there is a characteristic knee in the graph. If the expected raw BER has been misjudged, the consequences can be disastrous. Another result demonstrated by the example is that we can only guarantee to detect the same number of bits in error as there are redundant bits.

3.14 Block and convolutional codes

Figure 3.21(a) shows a strategy known as a crossword code, or product code. The data are formed into a two-dimensional array, in which each

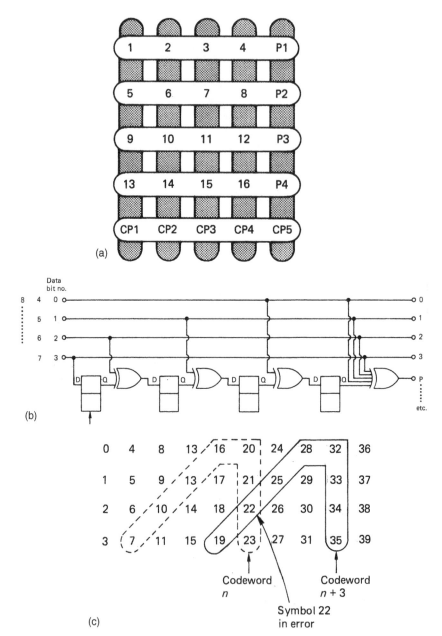

Figure 3.21   A block code is shown in (a). Each location in the block can be a bit or a word. Horizontal parity checks are made by adding P1, P2, etc., and cross-parity or vertical checks are made by adding CP1, CP2, etc. Any symbol in error will be at the intersection of the two failing codewords. In (b) a convolutional coder is shown. Symbols entering are subject to different delays which result in the codewords in (c) being calculated. These have a vertical part and a diagonal part. A symbol in error will be at the intersection of the diagonal part of one code and the vertical part of another.

location can be a single-bit or a multi-bit symbol. Parity is then generated on both rows and columns. If a single bit or symbol fails, one row parity check and one column parity check will fail, and the failure can be located at the intersection of the two failing checks. Although two symbols in error confuse this simple scheme, using more complex coding in a two-dimensional structure is very powerful, and further examples will be given throughout this chapter.

The example of Figure 3.21(a) assembles the data to be coded into a block of finite size and then each codeword is calculated by taking a different set of symbols. This should be contrasted with the operation of the circuit of (b). Here the data are not in a block, but form an endless stream. A shift register allows four symbols to be available simultaneously to the encoder. The action of the encoder depends upon the delays. When symbol 3 emerges from the first delay, it will be added (modulo-2) to symbol 6. When this sum emerges from the second delay, it will be added to symbol 9 and so on. The codeword produced is shown in (c) to be bent such that it has a vertical section and a diagonal section. Four symbols later the next codeword will be created one column further over in the data.

This is a convolutional code because the coder always takes parity on the same pattern of symbols convolved with the data stream on an endless basis. Figure 3.21(c) also shows that if an error occurs, it can be located because it will cause parity errors in two codewords. The error will be on the diagonal part of one codeword and on the vertical part of the other so that it can be located uniquely at the intersection and corrected by parity.

Comparison with the block code of Figure 3.21(a) will show that the convolutional code needs less redundancy for the same single-symbol location and correction performance as only a single redundant symbol is required for every four data symbols. Convolutional codes are computed on an endless basis which makes them inconvenient in networks where packet multiplexing is anticipated. Here the block code is more appropriate as it allows switching gaps to be created between codes.

Convolutional codes work best in channels suffering from Gaussian noise and will be found in broadcasting. They can easily be taken beyond their correcting power if used with a bursty channel.

3.15 Cyclic codes

In digital audio and video, relatively large quantities of data are involved and it is desirable to use relatively large data blocks to reduce the amount of bandwidth devoted to preambles, addressing and synchronizing. The principle of codewords having a special characteristic will still be employed, but they will be generated and checked algorithmically by equations. The syndrome will then be converted to the bit(s) in error by solving equations.

Where data can be accessed serially, simple circuitry can be used because the same gate will be used for many XOR operations. The circuit of Figure 3.22 is a kind of shift register, but with a particular feedback arrangement which leads it to be known as a twisted-ring counter. If seven message bits A–G are applied serially to this circuit, and each one of them is clocked, the outcome can be followed in the diagram. As bit A is presented and the system is clocked, bit A will enter the left-hand latch. When bits B and C are presented, A moves across to the right. Both XOR gates will have A on the upper input from the right-hand latch, the left one has D on the lower input and the right one has B on the lower input. When clocked, the left latch will thus be loaded with the XOR of A and D, and the right one with the XOR of A and B. The remainder of the sequence can be followed, bearing in mind that when the same term appears on both inputs of an XOR gate, it goes out, as the exclusive-OR of something

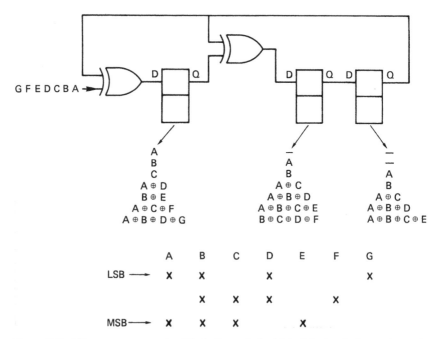

Figure 3.22 When seven successive bits A–G are clocked into this circuit, the contents of the three latches are shown for each clock. The final result is a parity-check matrix.

with itself is nothing. At the end of the process, the latches contain three different expressions. Essentially, the circuit makes three parity checks through the message, leaving the result of each in the three stages of the register. In the figure, these expressions have been used to draw up a check matrix. The significance of these steps can now be explained.

The bits A B C and D are four data bits, and the bits E F and G are redundancy. When the redundancy is calculated, bit E is chosen so that there are an even number of ones in bits A B C and E; bit F is chosen such that the same applies to bits B C D and F, and similarly for bit G. Thus the four data bits and the three check bits form a seven-bit codeword. If there is no error in the codeword, when it is fed into the circuit shown, the result of each of the three parity checks will be zero and every stage of the shift register will be cleared. As the register has eight possible states, and one of them is the error-free condition, then there are seven remaining states, hence the seven-bit codeword. If a bit in the codeword is corrupted, there will be a non-zero result. For example, if bit D fails, the check on bits A B D and G will fail, and a one will appear in the left-hand latch. The check on bits B C D F will also fail, and the centre latch will set. The check on bits A B C E will not fail, because D is not involved in it, making the right-hand bit zero. There will be a syndrome of 110 in the register, and this will be seen from the check matrix to correspond to an error in bit D. Whichever bit fails, there will be a different three-bit syndrome that uniquely identifies the failed bit. As there are only three latches, there can be eight different syndromes. One of these is zero, which is the error-free condition, and so there are seven remaining error syndromes. The length of the codeword cannot exceed seven bits, or there would not be enough syndromes to correct all the bits. This can also be made to tie in with the generation of the check matrix. If 14 bits, A to N, were fed into the circuit shown, the result would be that the check matrix repeated twice, and if a syndrome of 101 were to result, it could not be determined whether bit D or bit K failed. Because the check repeats every seven bits, the code is said to be a cyclic redundancy check (CRC) code.

It has been seen that the circuit shown makes a matrix check on a received word to determine if there has been an error, but the same circuit can also be used to generate the check bits. To visualize how this is done, examine what happens if only the data bits A B C and D are known, and the check bits E F and G are set to zero. If this message, ABCD000, is fed into the circuit, the left-hand latch will afterwards contain the XOR of A B C and zero, which is, of course, what E should be. The centre latch will contain the XOR of B C D and zero, which is what F should be and so on. This process is not quite ideal, however, because it is necessary to wait for three clock periods after entering the data before the check bits are available. Where the data are simultaneously being recorded and fed into the encoder, the delay would prevent the check bits being easily added to the end of the data stream. This problem can be overcome by slightly modifying the encoder circuit as shown in Figure 3.23. By moving the position of the input to the right, the operation of the circuit is advanced so that the check bits are ready after only four clocks. The process can be followed in the diagram for the four data bits A B C and D. On the first clock, bit A enters the left two latches, whereas on the second clock, bit B will appear on the upper input of the left XOR gate, with bit A on the lower input, causing the centre latch to load the XOR of A and B and so on.

The way in which the cyclic codes work has been described in engineering terms, but it can be described mathematically if analysis is contemplated.

Just as the position of a decimal digit in a number determines the power of ten (whether that digit means one, ten or 100), the position of a binary digit determines the power of two (whether it means one, two or four). It is possible to rewrite a binary number so that it is expressed as a list of powers of two. For example, the binary number 1101 means $8 + 4 + 1$, and can be written:

$$2^3 + 2^2 + 2^0$$

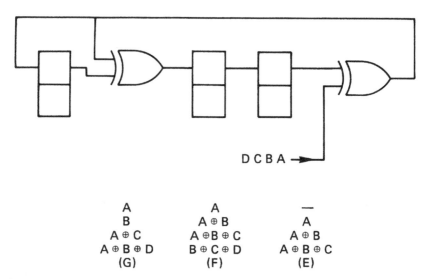

$$
\begin{array}{ccc}
A & A & - \\
B & A \oplus B & A \\
A \oplus C & A \oplus B \oplus C & A \oplus B \\
A \oplus B \oplus D & B \oplus C \oplus D & A \oplus B \oplus C \\
(G) & (F) & (E)
\end{array}
$$

Figure 3.23 By moving the insertion point three places to the right, the calculation of the check bits is completed in only four clock periods and they can follow the data immediately. This is equivalent to premultiplying the data by x^3.

In fact, much of the theory of error correction applies to symbols in number bases other than 2, so that the number can also be written more generally as

$$x^3 + x^2 + 1 \ (2^0 = 1)$$

and which also looks much more impressive. This expression, containing as it does various powers, is, of course, a polynomial. The circuit of Figure 3.22, which has been seen to construct a parity-check matrix on a codeword, can also be described as calculating the remainder due to dividing the input by a polynomial using modulo-2 arithmetic. In modulo-2 there are no borrows or carries, and addition and subtraction are replaced by the XOR function, which makes hardware implementation very easy. In Figure 3.24 it will be seen that the circuit of Figure 3.22 actually divides the codeword by the polynomial

$$x^3 + x + 1 \text{ or } 1011$$

This can be deduced from the fact that the right-hand bit is fed into two lower-order stages of the register at once. Once all the bits of the message have been clocked in, the circuit contains the remainder. In mathematical terms, the special property of a codeword is that it is a polynomial yielding a remainder of zero when divided by the generating polynomial. The receiver will make this division, and the result should be zero in the error-free case. Thus the codeword itself disappears from the division. If an error has occurred it is considered that this is due to an error polynomial added to the codeword polynomial. If a codeword divided by the check polynomial is zero, a non-zero syndrome must represent the error polynomial divided by the check polynomial. Thus if the syndrome is multiplied by the check polynomial, the latter will be cancelled out and the result will be the error polynomial. If this is added modulo-2 to the received word, it will cancel out the error and leave the corrected data.

Some examples of modulo-2 division are given in Figure 3.24, which can be compared with the parallel computation of parity checks according to the matrix of Figure 3.22.

The process of generating the codeword from the original data can also be described mathematically. If a codeword has to give zero remainder when divided, it follows that the data can be converted to a codeword by adding the remainder when the data are divided. Generally speaking, the remainder would have to be subtracted, but in modulo-2 there is no distinction. This process is also illustrated in Figure 3.24. The four data bits have three zeros placed on the right-hand end, to make the word length

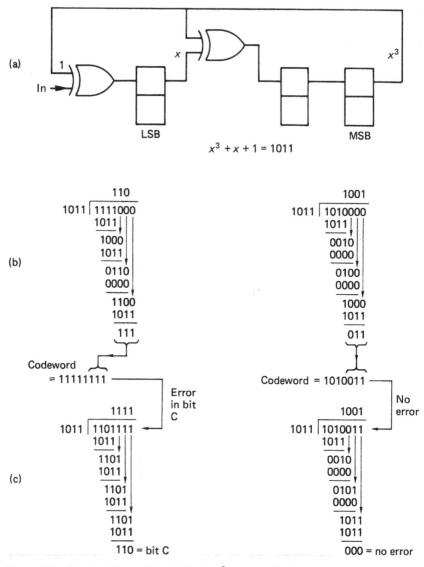

Figure 3.24 Circuit of Figure 3.22 divides by $x^3 + x + 1$ to find remainder. At (b) this is used to calculate check bits. At (c) right, zero syndrome, no error.

equal to that of a codeword, and this word is then divided by the polynomial to calculate the remainder. The remainder is added to the zero-extended data to form a codeword. The modified circuit of Figure 3.23 can be described as premultiplying the data by x^3 before dividing.

CRC codes are of primary importance for detecting errors, and several have been standardized for use in digital communications. The most common of these are:

$$x^{16} + x^{15} + x^2 + 1 \text{ (CRC-16)}$$

$$x^{16} + x^{12} + x^5 + 1 \text{ (CRC-CCITT)}$$

The 16-bit cyclic codes have codewords of length $2^{16}-1$ or 65 535 bits long. This may be too long for the application. Another problem with very long codes is that with a given raw BER, the longer the code, the more errors will occur in it. There may be enough errors to exceed the power of the code. The solution in both cases is to shorten or *puncture* the code. Figure 3.25 shows that in a punctured code, only the end of the codeword is used, and the data and redundancy are preceded by a string of zeros. It is not necessary to transmit these zeros, and, of course, errors cannot occur in them. Implementing a punctured code is easy. If a CRC generator starts with the register cleared and is fed with serial zeros, it will not change its state. Thus it is not necessary to provide the zeros, encoding can begin with the first data bit. In the same way, the leading zeros need not be provided during reception. The only precaution needed is that if a syndrome calculates the location of an error, this will be from the beginning of the codeword not from the beginning of the data. Where codes are used for detection only, this is of no consequence.

CRCs are in common use in digital interfaces. The channel status data of the AES/EBU interface, video frames in SD EDH and active lines in HD-SDI are all checked with CRCs.

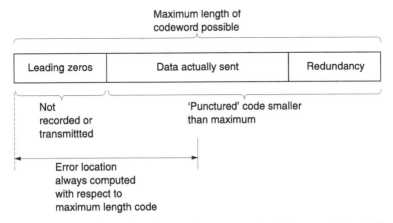

Figure 3.25 Codewords are often shortened, or punctured, which means that only the end of the codeword is actually transmitted. The only precaution to be taken when puncturing codes is that the computed position of an error will be from the beginning of the codeword, not from the beginning of the message.

3.16 The Galois field

Figure 3.26 shows a simple circuit consisting of three D-type latches which are clocked simultaneously. They are connected in series to form a shift register. At (a) a feedback connection has been taken from the output to the input and the result is a ring counter where the bits contained will recirculate endlessly. At (b) one XOR gate is added so that the output is fed back to more than one stage. The result is known as a twisted-ring counter and it has some interesting properties. Whenever the circuit is clocked, the left-hand bit moves to the right-hand latch, the centre bit moves to the left-hand latch and the centre latch becomes the XOR of the two outer latches. The figure shows that whatever the starting condition of the three bits in the latches, the same state will always be reached again after seven clocks, except if zero is used. The states of the latches form an endless ring of non-sequential numbers called a Galois field after the French mathematical prodigy Evariste Galois who discovered them. The states of the circuit form a maximum length sequence because there are as

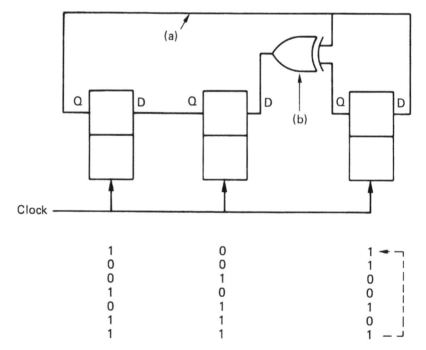

Figure 3.26 The circuit shown is a twisted-ring counter which has an unusual feedback arrangement. Clocking the counter causes it to pass through a series of non-sequential values. See text for details.

many states as are permitted by the word length. As the states of the sequence have many of the characteristics of random numbers, yet are repeatable, the result can also be called a pseudo-random sequence (prs). As the all-zeros case is disallowed, the length of a maximum length sequence generated by a register of m bits cannot exceed $(2m^{-1})$ states. The Galois field, however, includes the zero term. It is useful to explore the bizarre mathematics of Galois fields which use modulo-2 arithmetic. Familiarity with such manipulations is helpful when studying the error correction, particularly the Reed–Solomon codes used in recorders. They will also be found in processes which require pseudo-random numbers such as digital dither and randomized channel codes used in, for example, DVB.

The circuit of Figure 3.26 can be considered as a counter and the four points shown will then be representing different powers of 2 from the MSB on the left to the LSB on the right. The feedback connection from the MSB to the other stages means that whenever the MSB becomes 1, two other powers are also forced to one so that the code of 1011 is generated.

Each state of the circuit can be described by combinations of powers of x, such as

$$x^2 = 100$$
$$x = 010$$
$$x^2 + x = 110, \text{ etc.}$$

This fact that three bits have the same state because they are connected together is represented by the modulo-2 equation:

$$x^3 + x + 1 = 0$$

Let $x = a$, which is a primitive element. Now

$$a^3 + a + 1 = 0 \qquad\qquad (3.1)$$

In modulo-2

$$a + a = a^2 + a^2 = 0$$
$$a = x = 010$$
$$a^2 = x^2 = 100$$
$$a^3 = a + 1 = 011 \text{ from (3.1)}$$
$$a^4 = a \times a^3 = a(a + 1) = a^2 + a = 110$$

$$a^5 = a^2 + a + 1 = 111$$
$$a^6 = a \times a^5 = a(a^2 + a + 1)$$
$$= a^3 + a^2 + a = a + 1 + a^2 + a$$
$$= a^2 + 1 = 101$$
$$a^7 = a(a^2 + 1) = a^3 + a$$
$$= a + 1 + a = 1 = 001$$

In this way it can be seen that the complete set of elements of the Galois field can be expressed by successive powers of the primitive element. Note that the twisted-ring circuit of Figure 3.26 simply raises *a* to higher and higher powers as it is clocked. Thus the seemingly complex multi-bit changes caused by a single clock of the register become simple to calculate using the correct primitive and the appropriate power.

The numbers produced by the twisted-ring counter are not random; they are completely predictable if the equation is known. However, the sequences produced are sufficiently similar to random numbers that in many cases they will be useful. They are thus referred to as pseudo-random sequences. The feedback connection is chosen such that the expression it implements will not factorize. Otherwise a maximum-length sequence could not be generated because the circuit might sequence around one or other of the factors depending on the initial condition. A useful analogy is to compare the operation of a pair of meshed gears. If the gears have a number of teeth which is relatively prime, many revolutions are necessary to make the same pair of teeth touch again. If the number of teeth have a common multiple, far fewer turns are needed.

3.17   Introduction to the Reed–Solomon codes

The Reed–Solomon codes (Irving Reed and Gustave Solomon) are inherently burst correcting[4] because they work on multi-bit symbols rather than individual bits. The R–S codes are also extremely flexible in use. One code may be used both to detect and correct errors. The number of bursts to be corrected can be chosen at the design stage by the amount of redundancy. A further advantage of the R–S codes is that they can be used in conjunction with a separate error-detection mechanism in which case they perform the correction only by erasure. R–S codes operate at the theoretical limit of correcting efficiency. In other words, no more efficient code can be found.

In the simple CRC system described in section 3.15, the effect of the error is detected by ensuring that the codeword can be divided by a polynomial. The CRC codeword was created by addition of a redundant symbol to the data. In the Reed–Solomon codes, several errors can be isolated if the codeword is divisible by a number of polynomials. Clearly, if the codeword must divide by, say, two polynomials, it must have two redundant symbols. This is the minimum case of an R–S code. On receiving an R–S-coded message there will be two syndromes following the division. In the error-free case, these will both be zero. If both are not zero, there is an error.

It has been stated that the effect of an error is to add an error polynomial to the message polynomial. The number of terms in the error polynomial is the same as the number of errors in the codeword. The codeword divides to zero and the syndromes are a function of the error only. There are two syndromes and two equations. By solving these simultaneous equations it is possible to obtain two unknowns. One of these is the position of the error, known as the *locator,* and the other is the error bit pattern, known as the *corrector.* As the locator is the same size as the code symbol, the length of the codeword is determined by the size of the symbol. A symbol size of eight bits is commonly used because it fits in conveniently with both 16-bit audio samples and byte-oriented computers. An eight-bit syndrome results in a locator of the same word length. Eight bits have 2^8 combinations, but one of these is the error-free condition, and so the locator can specify one of only 255 symbols. As each symbol contains eight bits, the codeword will be $255 \times 8 = 2040$ bits long.

As further examples, five-bit symbols could be used to form a codeword 31 symbols long, and three-bit symbols would form a codeword seven symbols long. This latter size is small enough to permit some worked examples, and will be used further here. Figure 3.27 shows that in the seven-symbol codeword, five symbols of three bits each, A–E, are the data, and P and Q are the two redundant symbols. This simple example will locate and correct a single symbol in error. It does not matter, however, how many bits in the symbol are in error.

The two check symbols are solutions to the following equations:

$$A \oplus B \oplus C \oplus D \oplus E \oplus P \oplus Q = 0 \ (\oplus = \text{XOR symbol})$$
$$a^7 A \oplus a^6 B \oplus a^5 C \oplus a^4 D \oplus a^3 E \oplus a^2 P \oplus aQ = 0$$

where a is a constant. The original data A–E followed by the redundancy P and Q pass through the channel.

The receiver makes two checks on the message to see if it is a codeword. This is done by calculating syndromes using the following

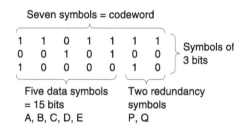

Seven symbols = codeword

Five data symbols
= 15 bits
A, B, C, D, E

Two redundancy
symbols
P, Q

Symbols of
3 bits

Figure 3.27   A Reed–Solomon codeword. As the symbols are of three bits, there can only be eight possible syndrome values. One of these is all zeros, the error-free case, and so it is only possible to point to seven errors: hence the codeword length of seven symbols. Two of these are redundant, leaving five data symbols.

expressions, where the (′) implies the received symbol which is not necessarily correct:

$$S_0 = A' \oplus B' \oplus C' \oplus D' \oplus E' \, P' \oplus Q'$$

(This is in fact a simple parity check.)

$$S_1 = a^7 A' \oplus a^6 B' \oplus a^5 C' \oplus a^4 D' \oplus a^3 E' \oplus a^2 P' \oplus aQ'$$

If two syndromes of all zeros are not obtained, there has been an error. The information carried in the syndromes will be used to correct the error. For the purpose of illustration, let it be considered that D′ has been corrupted before moving to the general case. D′ can be considered to be the result of adding an error of value E to the original value D such that D′ = D $\oplus$ E.

As A $\oplus$ B $\oplus$ C $\oplus$ D $\oplus$ E $\oplus$ P $\oplus$ Q = 0
then A $\oplus$ B $\oplus$ C $\oplus$ (D $\oplus$ E) $\oplus$ E $\oplus$ P $\oplus$ Q = E = S_0
As D′ = D $\oplus$ E
then D = D′ $\oplus$ E = D′ $\oplus$ S_0

Thus the value of the corrector is known immediately because it is the same as the parity syndrome S_0. The corrected data symbol is obtained simply by adding S_0 to the incorrect symbol.

At this stage, however, the corrupted symbol has not yet been identified, but this is equally straightforward:

As a^7 A $\oplus$ a^6 B $\oplus$ a^5 C $\oplus$ a^4 D $\oplus$ a^3 E $\oplus$ a^2 P $\oplus$ aQ = 0 then:

$$a^7 \text{ A} \oplus a^6 \text{ B} \oplus a^5 \text{ C} \oplus a^4 \text{ (D} \oplus \text{E)} \oplus a^3 \text{ E} \oplus a^2 \text{ P} \oplus a\text{Q} = a^4 \text{ } E = S_1$$

Thus the syndrome S_1 is the error bit pattern E, but it has been raised to a power of a that is a function of the position of the error symbol in the block. If the position of the error is in symbol k, then k is the locator value and:

$$S_0 \times a^k = S_1$$

Hence:

$$a^k = \frac{S_1}{S_0}$$

The value of k can be found by multiplying S_0 by various powers of a until the product is the same as S_1. Then the power of a necessary is equal to k. The use of the descending powers of a in the codeword calculation is now clear because the error is then multiplied by a different power of a dependent upon its position, known as the locator, because it gives the position of the error. The process of finding the error position by experiment is known as a Chien search.

Whilst the expressions above show that the values of P and Q are such that the two syndrome expressions sum to zero, it is not yet clear how P and Q are calculated from the data. Expressions for P and Q can be found by solving the two R–S equations simultaneously. This has been done in Appendix 3.1. The following expressions must be used to calculate P and Q from the data in order to satisfy the codeword equations. These are:

$$P = a^6 \text{ A} \oplus a\text{B} \oplus a^2 \text{ C} \oplus a^5 \text{ D} \oplus a^3 \text{ E}$$
$$Q = a^2 \text{ A} \oplus a^3 \text{ B} \oplus a^6 \text{ C} \oplus a^4 \text{ D} \oplus a\text{E}$$

In both the calculation of the redundancy shown here and the calculation of the corrector and the locator it is necessary to perform numerous multiplications and raising to powers. This appears to present a formidable calculation problem at both the encoder and the decoder. This would be the case if the calculations involved were conventionally executed. However, the calculations can be simplified by using logarithms. Instead of multiplying two numbers, their logarithms are added. In order to find the cube of a number, its logarithm is added three times. Division is performed by subtraction of the logarithms. Thus all the manipulations necessary can be achieved with addition or subtraction, which is straightforward in logic circuits.

The success of this approach depends upon simple implementation of log tables. Raising a constant, a, known as the *primitive element*, to successively higher powers in modulo-2 gives rise to a Galois field. Each element of the field represents a different power n of a. It is a fundamental of the R–S codes that all the symbols used for data, redundancy and syndromes are considered to be elements of a Galois field. The number of bits in the symbol determines the size of the Galois field, and hence the number of symbols in the codeword.

In Figure 3.28, the binary values of the elements are shown alongside the power of a they represent. In the R–S codes, symbols are no longer considered simply as binary numbers, but also as equivalent powers of a. In Reed–Solomon coding and decoding, each symbol will be multiplied by some power of a. Thus if the symbol is also known as a power of a it is only necessary to add the two powers. For example, if it is necessary to multiply the data symbol 100 by a^3, the calculation proceeds as follows, referring to Figure 3.28:

$$100 = a^2 \text{ so } 100 \times a^3 = a^{(2+3)} = a^5 = 111$$

Note that the results of a Galois multiplication are quite different from binary multiplication. Because all products must be elements of the field,

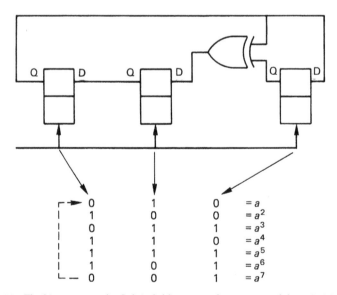

Figure 3.28 The bit patterns of a Galois field expressed as powers of the primitive element a. This diagram can be used as a form of log table in order to multiply binary numbers. Instead of an actual multiplication, the appropriate powers of a are simply added.

sums of powers that exceed seven wrap around by having seven subtracted. For example:

$$a^5 \times a^6 = a^{11} = a^4 = 110$$

Figure 3.29 gives an example of the Reed–Solomon encoding process.

The Galois field shown in Figure 3.28 has been used, having the primitive element $a = 010$. At the beginning of the calculation of P, the symbol A is multiplied by a^6. This is done by converting A to a power of a. According to Figure 3.28, $101 = a^6$ and so the product will be $a^{(6 + 6)} = a^{12} = a^5 = 111$. In the same way, B is multiplied by a, and so on, and the products are added modulo-2. A similar process is used to calculate Q.

Figure 3.29 also demonstrates that the codeword satisfies the checking equations. The modulo-2 sum of the seven symbols, S_0, is 000 because each column has an even number of ones. The calculation of S_1 requires multiplication by descending powers of a. The modulo-2 sum of the products is again zero. These calculations confirm that the redundancy calculation was properly carried out.

Figure 3.30 gives three examples of error correction based on this codeword. The erroneous symbol is marked with a dash. As there has been an error, the syndromes S_0 and S_1 will not be zero.

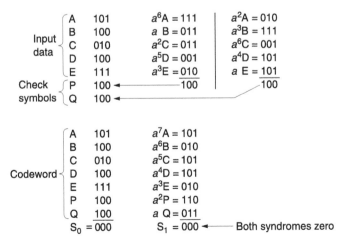

Figure 3.29 Five data symbols A–E are used as terms in the generator polynomials derived in Appendix 3.1 to calculate two redundant symbols P and Q. An example is shown at the top. Below is the result of using the codeword symbols A–Q as terms in the checking polynomials. As there is no error, both syndromes are zero.

7	A	101
6	B	100
5	C	010
4	D'	101
3	E	111
2	P	100
1	Q	100
	S_0 =	001

$a^7 A = 101$
$a^6 B = 010$
$a^5 C = 101$
$a^4 D' = 011$
$a^3 E = 010$
$a^2 P = 110$
$a\, Q = 011$
$S_1 = 110$

$$\frac{S_1}{S_0} = \frac{a^4}{1} = a^4$$

$k = 4$

$D' + S_0 = 101 + 001$
$D = 100$

7	A	101
6	B	100
5	C'	110
4	D	100
3	E	111
2	P	100
1	Q	100
	S_0 =	100

$a^7 A = 101$
$a^6 B = 010$
$a^5 C = 100$
$a^4 D = 101$
$a^3 E = 010$
$a^2 P = 110$
$a\, Q = 011$
$S_1 = 001$

$$\frac{S_1}{S_0} = \frac{1}{a^2} = \frac{1}{a^2} \times \frac{a^5}{a^5} = a^5$$

$k = 5$

$C' + S_0 = 110 + 100$
$C = 010$

7	A'	111
6	B	100
5	C	010
4	D	100
3	E	111
2	P	100
1	Q	100
	S_0 =	010

$a^7 A = 111$
$a^6 B = 010$
$a^5 C = 101$
$a^4 D = 101$
$a^3 E = 010$
$a^2 P = 110$
$a\, Q = 011$
$S_1 = 010$

$$\frac{S_1}{S_0} = \frac{a}{a} = 001 = a^7$$

$k = 7$

$A' + S_0 = 111 + 010$
$A = 101$

Figure 3.30 Three examples of error location and correction. The number of bits in error in a symbol is irrelevant; if all three were wrong, S_0 would be 111, but correction is still possible.

3.18 Correction by erasure

In the examples of Figure 3.30, two redundant symbols P and Q have been used to locate and correct one error symbol. If the positions of errors are known by some separate mechanism (see product codes, section 3.19) the locator need not be calculated. The simultaneous equations may instead be solved for two correctors. In this case the number of symbols that can be corrected is equal to the number of redundant symbols. In Figure 3.31(a) two errors have taken place, and it is known that they are in symbols C and D. Since S_0 is a simple parity check, it will reflect the modulo-2 sum of the two errors. Hence $S_1 = EC \oplus ED$.

The two errors will have been multiplied by different powers in S_1, such that:

$$S_1 = a^5\, EC \oplus a^4\, ED$$

$$
\begin{array}{llll}
A & 1\underline{0}1 & a^7 A = & 101 \\
B & 100 & a^6 B = & 010 \\
(C \oplus E_C) & 001 & a^5 (C \oplus E_C) & 111 \\
(D \oplus E_D) & 0\,1\,0 & a^4 (D \oplus E_D) & 111 \\
E & 111 & a^3 E = & 010 \\
P & 100 & a^2 P = & 110 \\
Q & 100 & a\,Q = & 011 \\
S_1 \;=\; & 1\overline{01} & S_1 \;=\; & \overline{000}
\end{array}
$$

$$S_0 = E_C \oplus E_D \qquad\qquad S_1 = a^5 E_C \oplus a^4 E_D$$

$$S_1 = a^5 E_C \oplus a^4 (S_0 \oplus E_C)$$

$$= a^5 E_C \oplus a^4 S_0 \oplus a^4 E_C$$

$$\therefore E_C = \frac{S_1 \oplus a^4 S_0}{a^5 \oplus a^4} = \frac{000 \oplus 011}{001} = 011$$

$$C = (C \oplus E_C) \oplus E_C = 001 \oplus 011 = \underline{010}$$

$$S_1 = a^5 (S_0 \ominus E_D) \oplus a^4 E_D$$

$$= a^5 S_0 \oplus a^5 E_D \oplus a^4 E_D$$

$$\therefore E_D = \frac{S_1 \oplus a^5 S_0}{a^5 \oplus a^4} = \frac{000 \oplus 110}{001} = 110$$

$$D = (D \oplus E_D) + E_D = 010 \oplus 110 = \underline{100} \qquad\qquad \textbf{(a)}$$

$$
\begin{array}{lll}
A & 101 & a^7 A = 101 \\
B & 100 & a^6 B = 010 \quad S_0 = C \oplus D \\
C & \underline{000} & a^5 C = \underline{000} \\
D & \underline{000} & a^4 D = \underline{000} \quad S_1 = a^5 C \oplus a^4 D \\
E & 111 & a^3 E = 010 \\
P & 100 & a^2 P = 110 \\
Q & 100 & a\,Q = 011 \\
S_0 & = 100 & S_1 \;= 000
\end{array}
$$

$$S_1 = a^5 S_0 \oplus a^5 D \oplus a^4 D = a^5 S_0 \oplus D$$

$$\therefore D = S_1 \oplus a^5 S_0 = 000 \oplus 100 = \underline{100}$$

$$S_1 = a^5 C \oplus a^4 C \oplus a^4 S_0 = C \oplus a^4 S_0$$

$$\therefore C = S_1 \oplus a^4 S_0 = 000 \oplus 010 = \underline{010}$$

$$\textbf{(b)}$$

Figure 3.31 If locations of errors are known, the syndromes are a known function of the two errors as shown in (a). It is, however, much simpler to set the incorrect symbols to zero, that is, to *erase* them as in (b). Then the syndromes are a function of the wanted symbols and correction is easier.

These two equations can be solved, as shown in the figure, to find *EC* and *ED*, and the correct value of the symbols will be obtained by adding these correctors to the erroneous values. It is, however, easier to set the values of the symbols in error to zero. In this way the nature of the error is rendered irrelevant and it does not enter the calculation. This setting of symbols to zero gives rise to the term erasure. In this case,

$$S_0 = C \oplus D$$

$$S_1 = a^5 C + a^4 D$$

Erasing the symbols in error makes the errors equal to the correct symbol values and these are found more simply as shown in Figure 3.30 (b).

Practical systems will be designed to correct more symbols in error than in the simple examples given here. If it is proposed to correct by erasure an arbitrary number of symbols in error given by t, the codeword must be divisible by t different polynomials. Alternatively, if the errors must be located and corrected, $2t$ polynomials will be needed. These will be of the form $(x + a^n)$ where n takes all values up to t or $2t$, where a is the primitive element.

Where four symbols are to be corrected by erasure, or two symbols are to be located and corrected, four redundant symbols are necessary, and the codeword polynomial must then be divisible by

$$(x + a^0) (x + a^1) (x + a^2) (x + a^3)$$

Upon receipt of the message, four syndromes must be calculated, and the four correctors or the two error patterns and their positions are determined by solving four simultaneous equations. This generally requires an iterative procedure, and a number of algorithms have been developed for the purpose.[5-7] Modern FEC systems, such as ATM and DVB, use eight-bit R–S codes and erasure extensively. The primitive polynomial commonly used with GF(256) is:

$$x^8 + x^4 + x^3 + x^2 + 1$$

The codeword will be 255 bytes long but will often be shortened by puncturing. The larger Galois fields require less redundancy, but the computational problem increases. LSI chips have been developed specifically for R–S decoding in many high-volume formats.

3.19 Interleaving

The concept of bit interleaving was introduced in connection with a single-bit correcting code to allow it to correct small bursts. With burst-correcting codes such as Reed–Solomon, bit interleave is unnecessary. In some channels, the burst size may be many bytes rather than bits, and to rely on a code alone to correct such errors would require a lot of redundancy. The solution in this case is to employ symbol interleaving, as shown in Figure 3.32. Several codewords are encoded from input data, but these are not recorded in the order they were input, but are physically reordered in the channel, so that a real burst error is split into smaller bursts in several codewords. The size of the burst seen by each codeword is now determined primarily by the parameters of the interleave, and Figure 3.33 shows that the probability of occurrence of bursts with respect to the burst length in a given codeword is modified. The number of bits in the interleave word can be made equal to the burst-correcting ability of the code in the knowledge that it will be exceeded only very infrequently.

There are a number of different ways in which interleaving can be performed. Figure 3.34 shows that, in block interleaving, words are reordered within blocks that are correctly ordered. The block interleave is achieved by writing samples into a memory in sequential address locations from a counter, and reading the memory with non-sequential addresses from a sequencer. The effect is to convert a one-dimensional sequence of samples into a two-dimensional structure having rows and columns.

The alternative to block interleaving is convolutional interleaving where the interleave process is endless. In Figure 3.35 symbols are assembled into short blocks and then delayed by an amount proportional to the position in the block. It will be seen from the figure that the delays have the effect of shearing the symbols so that columns on the left

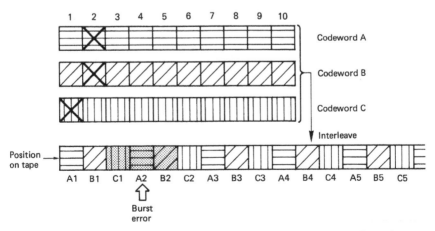

Figure 3.32 The interleave controls the size of burst errors in individual codewords.

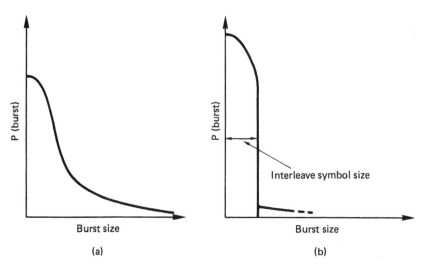

Figure 3.33   (a) The distribution of burst sizes might look like this. (b) Following interleave, the burst size within a codeword is controlled to that of the interleave symbol size, except for gross errors which have low probability.

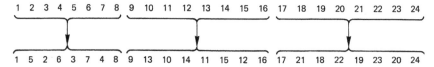

Figure 3.34   In block interleaving, data are scrambled within blocks that are themselves in the correct order.

side of the diagram become diagonals on the right. When the columns on the right are read, the convolutional interleave will be obtained. Convolutional interleave works well in transmission applications such as DVB where the signal is continuous. Convolutional interleave has the advantage of requiring less memory to implement than a block code. This is because a block code requires the entire block to be written into the memory before it can be read, whereas a convolutional code requires only enough memory to cause the required delays.

3.20 Product codes

In the presence of burst errors alone, the system of interleaving works very well, but it is known that in most practical channels there are also uncorrelated errors of a few bits due to noise. Figure 3.36 shows an

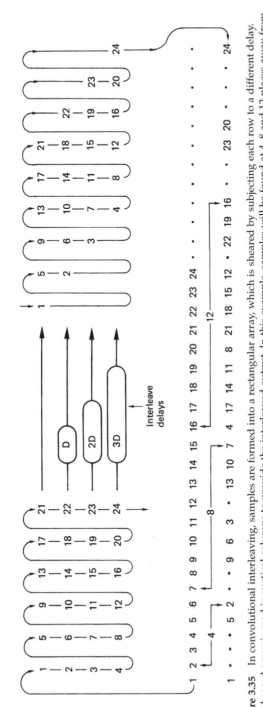

Figure 3.35 In convolutional interleaving, samples are formed into a rectangular array, which is sheared by subjecting each row to a different delay. The sheared array is read in vertical columns to provide the interleaved output. In this example, samples will be found at 4, 8 and 12 places away from their original order.

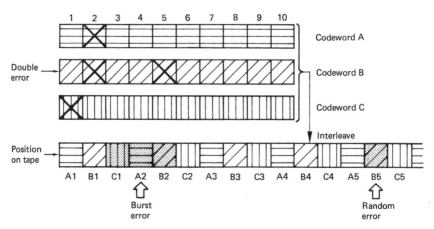

Figure 3.36 The interleave system falls down when a random error occurs adjacent to a burst.

interleaving system where a dropout-induced burst error has occurred which is at the maximum correctable size. All three codewords involved are working at their limit of one symbol. A random error due to noise in the vicinity of a burst error will cause the correction power of the code to be exceeded. Thus a random error of a single bit causes a further entire symbol to fail. This is a weakness of an interleaving system designed solely to handle dropout-induced bursts. Practical high-density equipment must address the problem of noise-induced or random errors and burst errors occurring at the same time. This is done by forming codewords both before and after the interleave process. In block interleaving, this results in a *product code*, whereas in the case of convolutional interleave the result is called *cross-interleaving*.

Figure 3.37 shows that in a product code the redundancy calculated first and checked last is called the outer code, and the redundancy calculated second and checked first is called the inner code. The inner code is formed along tracks on the medium. Random errors due to noise are corrected by the inner code and do not impair the burst-correcting power of the outer code. Burst errors are declared uncorrectable by the inner code which flags the bad samples on the way into the deinterleave memory. The outer code reads the error flags in order to correct the flagged symbols by erasure. The error flags are also known as erasure flags. As it does not have to compute the error locations, the outer code needs half as much redundancy for the same correction power. Thus the inner code redundancy does not raise the code overhead. The combination of codewords

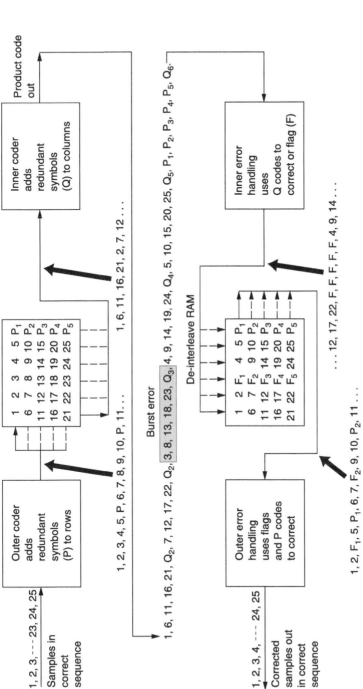

Figure 3.37 In addition to the redundancy P on rows, inner redundancy Q is also generated on columns. On replay, the Q code checker will pass on flags F if it finds an error too large to handle itself. The flags pass through the de-interleave process and are used by the outer error correction to identify which symbol in the row needs correcting with P redundancy. The concept of crossing two codes in this way is called a product code.

with interleaving in several dimensions yields a truly synergistic error-protection strategy, in that the end result is more powerful than the sum of the parts. This technique is used in the FEC strategy for delivering audio over ATM.

3.21   Networks

A network is basically a communication resource shared for economic reasons. Like any shared resource, decisions have to be made, somewhere and somehow, how the resource is to be used. In the absence of such decisions the resultant chaos will be such that the resource might as well not exist. In communications networks the resource is the ability to convey data from any node or port to any other. On a particular cable, clearly only one transaction of this kind can take place at any one instant even though in practice many nodes will simultaneously be wanting to transmit data. Arbitration is needed to determine which node is allowed to transmit.

There are a number of different arbitration protocols and these have evolved to support the needs of different types of network. In small networks, such as LANs, a single point failure halting the entire network may be acceptable, whereas in a public transport network owned by a telecommunications company, the network will be redundant so that if a particular link fails data may be sent via an alternative route. A link that has reached its maximum capacity may also be supplanted by transmission over alternative routes.

In physically small networks, arbitration may be carried out in a single location. This is fast and efficient, but if the arbitrator fails it leaves the system completely crippled. The processor buses in computers work in this way. In centrally arbitrated systems the arbitrator needs to know the structure of the system and the status of all the nodes. Following a configuration change, due perhaps to the installation of new equipment, the arbitrator needs to be told what the new configuration is, or have a mechanism which allows it to explore the network and learn the configuration. Central arbitration is only suitable for small networks that change their configuration infrequently.

In other networks the arbitration is distributed so that some decision-making ability exists in every node. This is less efficient but it does allow at least some of the network to continue operating after a component failure. Distributed arbitration also means that each node is self-sufficient and so no changes need to be made if the network is reconfigured by adding or deleting a node. This is the only possible approach in wide area networks where the structure may be very complex and change dynamically in the event of failures or overload.

Ethernet uses distributed arbitration. FireWire is capable of using both types of arbitration. A small amount of decision-making ability is built into every node so that distributed arbitration is possible. However, if one of the nodes happens to be a computer, it can run a centralized arbitration algorithm.

The physical structure of a network is subject to some variation as Figure 3.38 shows. In radial networks (a) each port has a unique cable connection to a device called a *hub*. The hub must have one connection for every port and this limits the number of ports. However, a cable failure will only result in the loss of one port. In a ring system (b) the nodes are connected like a daisy chain with each node acting as a feedthrough. In this case the arbitration requirement must be distributed. With some protocols, a single cable break doesn't stop the network operating. Depending on the protocol, simultaneous transactions may be possible provided they don't require the same cable. For example, in a storage network, a disk drive may be outputting data to an editor whilst another drive is backing up data to a tape streamer. For the lowest cost, all nodes are physically connected in parallel to the same cable. Figure 3.38(c) shows that a cable break would divide the network into two halves, but

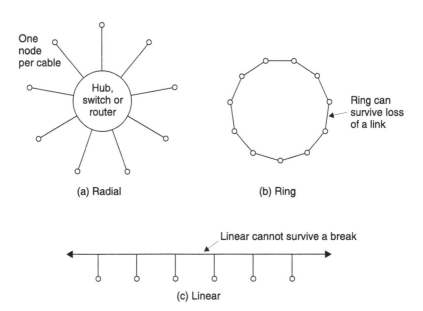

Figure 3.38 Network configurations. At (a) the radial system uses one cable to each node. (b) Ring system uses less cable than radial. (c) Linear system is simple but has no redundancy.

it is possible that the impedance mismatch at the break could stop both halves working.

One of the concepts involved in arbitration is priority, which is fundamental to providing an appropriate quality of service. If two processes both want to use a network, the one with the highest priority would normally go first. Attributing priority must be done carefully because some of the results are non-intuitive. For example, it may be beneficial to give a high priority to a humble device that has a low data rate for the simple reason that if it is given use of the network it won't need it for long. In a television environment transactions concerned with on-air processes would have priority over file transfers concerning production and editing.

When a device gains access to the network to perform a transaction, generally no other transaction can take place until it has finished. Consequently it is important to limit the amount of time that a given port can stay on the bus. In this way when the time limit expires, a further arbitration must take place. The result is that the network resource rotates between transactions rather than one transfer hogging the resource and shutting everyone else out.

It follows from the presence of a time (or data quantity) limit that ports must have the means to break large files up into frames or cells and reassemble them on reception. This process is sometimes called *adaptation*. If the data to be sent originally exist at a fixed bit rate, some buffering will be needed so that the data can be time-compressed into the available

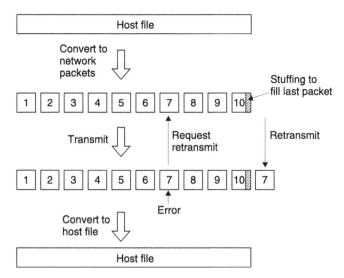

Figure 3.39 Receiving a file which has been divided into packets allows for the retransmission of just the packet in error.

frames. Each frame must be contiguously numbered and the system must transmit a file size or word count so that the receiving node knows when it has received every frame in the file.

The error-detection system interacts with this process because if any frame is in error on reception, the receiving node can ask for a retransmission of the frame. This is more efficient than retransmitting the whole file. Figure 3.39 shows the flow chart for a receiving node.

Breaking files into frames helps to keep down the delay experienced by each process using the network. Figure 3.40 shows that each frame may be stored ready for transmission in a silo memory. It is possible to make the priority a function of the number of frames in the silo, as this is a direct measure of how long a process has been kept waiting. Isochronous

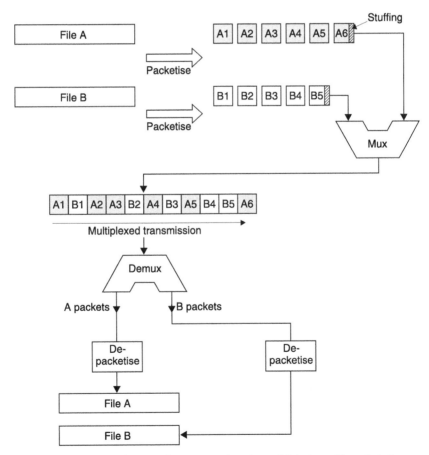

Figure 3.40 Files are broken into frames or packets for multiplexing with packets from other users. Short packets minimize the time between the arrival of successive packets. The priority of the multiplexing must favour isochronous data over asynchronous data.

systems must do this in order to meet maximum delay specifications. In Figure 3.40 once frame transmission has completed, the arbitrator will determine which process sends a frame next by examining the depth of all the frame buffers. MPEG transport stream multiplexers and networks delivering MPEG data must work in this way because the transfer is isochronous and the amount of buffering in a decoder is limited for economic reasons.

A central arbitrator is relatively simple to implement because when all decisions are taken centrally there can be no timing difficulty (assuming a well-engineered system). In a distributed system, there is an extra difficulty due to the finite time taken for signals to travel down the data paths between nodes.

Ethernet uses a protocol called CSMA/CD (carrier sense multiple access with collision detect) developed by DEC and Xerox. This is a distributed arbitration network where each node follows some simple rules. The first of these is not to transmit if an existing bus signal is detected. The second is not to transmit more than a certain quantity of data before releasing the bus. Devices wanting to use the bus will see bus signals and so will wait until the present bus transaction finishes. This must happen at some point because of the frame size limit. When the frame is completed, signalling on the bus should cease. The first device to sense the bus becoming free and to assert its own signal will prevent any other nodes transmitting according to the first rule. Where numerous devices are present it is possible to give them a priority structure by providing a delay between sensing the bus coming free and beginning a transaction. High-priority devices will have a short delay so they get in first. Lower-priority devices will only be able to start a transaction if the high-priority devices don't need to transfer.

It might be thought that these rules would be enough and everything would be fine. Unfortunately the finite signal speed means that there is a flaw in the system. Figure 3.41 shows why. Device A is transmitting and devices B and C both want to transmit and have equal priority. At the end of A's transaction, devices B and C see the bus become free at the same instant and start a transaction. With two devices driving the bus, the resultant waveform is meaningless. This is known as a collision and all nodes must have means to recover from it. First, each node will read the bus signal at all times. When a node drives the bus, it will also read back the bus signal and compare it with what was sent. Clearly if the two are the same all is well, but if there is a difference, this must be because a collision has occurred and two devices are trying to determine the bus voltage at once.

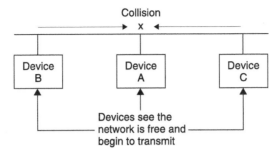

Figure 3.41 In Ethernet collisions can occur because of the finite speed of the signals. A 'back-off' algorithm handles collisions, but they do reduce the network throughput.

If a collision is detected, both colliding devices will sense the disparity between the transmitted and readback signals, and both will release the bus to terminate the collision. However, there is no point is adhering to the simple protocol to reconnect because this will simply result in another collision. Instead each device has a built-in delay that must expire before another attempt is made to transmit. This delay is not fixed, but is controlled by a random number generator and so changes from transaction to transaction.

The probability of two node devices arriving at the same delay is infinitesimally small. Consequently if a collision does occur, both devices will drop the bus, and they will start their back-off timers. When the first timer expires, that device will transmit and the other will see the transmission and remain silent. In this way the collision is not only handled, but also prevented from happening again.

The performance of Ethernet is usually specified in terms of the bit rate at which the cabling runs. However, this rate is academic because it is not available all the time. In a real network bit rate is lost by the need to send headers and error-correction codes and by the loss of time due to interframe spaces and collision handling. As the demand goes up, the number of collisions increases and throughput goes down. Collision-based arbitrators do not handle congestion well.

An alternative method of arbitration developed by IBM is shown in Figure 3.42. This is known as a *token ring* system. All the nodes have an input and an output and are connected in a ring that must be intact for the system to work. Data circulate in one direction only. If data are not addressed to a node that receives them, the data will be passed on. When the data arrive at the addressed node, that node will capture the data as well as passing them on with an acknowledge added. Thus the data packet travels right around the ring back to the sending node. When the sending node receives the acknowledge, it will transmit a token packet.

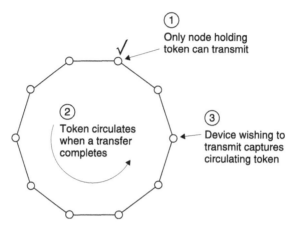

Figure 3.42 In a token ring system only the node in possession of the token can transmit so collisions are impossible. In very large rings the token circulation time causes loss of throughput.

This token packet passes to the next node, which will pass it on if it does not wish to transmit. If no device wishes to transmit, the token will circulate endlessly. However, if a device has data to send, it simply waits until the token arrives again and captures it. This node can now transmit data in the knowledge that there cannot be a collision because no other node has the token.

In simple token ring systems, the transmitting node transmits idle characters after the data packet has been sent in order to maintain synchronization. The idle character transmission will continue until the acknowledge arrives. In the case of long packets the acknowledge will arrive before the packet has all been sent and no idle characters are necessary. However, with short packets idle characters will be generated. These idle characters use up ring bandwidth.

Later token ring systems use early token release (ETR). After the packet has been transmitted, the sending node sends a token straight away. Another node wishing to transmit can do so as soon as the current packet has passed.

It might be thought that the nodes on the ring would transmit in their physical order, but this is not the case because a priority system exists. Each node can have a different priority if necessary. If a high-priority node wishes to transmit, as a packet from elsewhere passes through that node, the node will set *reservation bits* with its own priority level. When the sending node finishes and transmits a token, it will copy that priority level into the token. In this way nodes with a lower priority level will pass the token on instead of capturing it. The token will ultimately arrive at the high-priority node.

The token ring system has the advantage that it does not waste throughput with collisions and so the full capacity is always available. However, if the ring is broken the entire network fails.

In Ethernet the performance is degraded by number of transactions, not by the number of nodes, whereas in token ring, performance is degraded by the number of nodes.

3.22 MPEG packets and time stamps

The video elementary stream is an endless bitstream representing pictures that take a variable length of time to transmit. Bi-direction coding means that pictures are not necessarily in the correct order. Storage and transmission systems prefer discrete blocks of data and so elementary streams are packetized to form a PES (packetized elementary stream). Audio elementary streams are also packetized. A packet is shown in Figure 3.43. It begins with a header containing a unique packet start code and a code which identifies the type of data stream. Optionally the packet header also may contain one or more *time stamps* used for synchronizing the video decoder to real time and for obtaining lip sync.

Figure 3.44 shows that a time stamp is a sample of the state of a counter driven by a 90 kHz clock. This is obtained by dividing down the master 27 MHz clock of MPEG-2. This 27 MHz clock must be locked to the video frame rate and the audio sampling rate of the program concerned. There are two types of time stamp: PTS and DTS. These are abbreviations for presentation time stamp and decode time stamp. A presentation time stamp determines when the associated picture should be displayed on

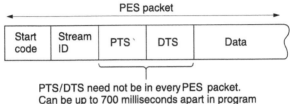

Figure 3.43 Program specific information helps the demultiplexer to select the required program.

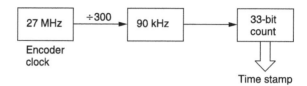

Figure 3.44   Time stamps are the result of sampling a counter driven by the encoder clock.

the screen, whereas a decode time stamp determines when it should be decoded. In bi-directional coding these times can be quite different.

Audio packets have only presentation time stamps. Clearly if lip sync is to be obtained, the audio sampling rate of a given program must have been locked to the same master 27 MHz clock as the video and the time stamps must have come from the same counter driven by that clock.

In practice, the time between input pictures is constant and so there is a certain amount of redundancy in the time stamps. Consequently PTS/DTS need not appear in every PES packet. Time stamps can be up to 100 ms apart in transport streams. As each picture type (*I*, *P* or *B*) is flagged in the bitstream, the decoder can infer the PTS/DTS for every picture from the ones actually transmitted.

The MPEG-2 transport stream is intended to be a multiplex of many TV programs with their associated sound and data channels, although a single program transport stream (SPTS) is possible. The transport stream is based upon packets of constant size so that multiplexing, adding error-correction codes and interleaving in a higher layer is eased. Figure 3.45 shows these to be always 188 bytes long.

Transport stream packets always begin with a header. The remainder of the packet carries data known as the payload. For efficiency, the normal header is relatively small, but for special purposes the header may be extended. In this case the payload gets smaller so that the overall size of the packet is unchanged. Transport stream packets should not be confused with PES packets that are larger and vary in size. PES packets are broken up to form the payload of the transport stream packets.

The header begins with a sync byte which is a unique pattern detected by a demultiplexer. A transport stream may contain many different elementary streams and these are identified by giving each a unique 13-bit packet identification code or PID which is included in the header. A multiplexer seeking a particular elementary stream simply checks the PID of every packet and accepts only those that are matching.

In a multiplex there may be many packets from other programs in between packets of a given PID. To help the demultiplexer, the packet

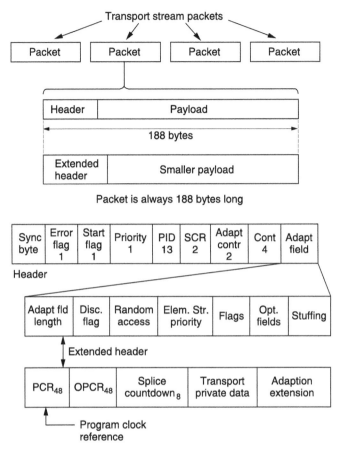

Figure 3.45 Transport stream packets are always 188 bytes long to facilitate multiplexing and error correction.

header contains a continuity count. This is a four-bit value which increments at each new packet having a given PID.

This approach allows statistical multiplexing as it does matter how many or how few packets have a given PID; the demux will still find them. Statistical multiplexing has the problem that it is virtually impossible to make the sum of the input bit rates constant. Instead the multiplexer aims to make the average data bit rate slightly less than the maximum and the overall bit rate is kept constant by adding 'stuffing' or null packets. These packets have no meaning, but simply keep the bit rate constant. Null packets always have a PID of 8191 (all ones) and the demultiplexer discards them.

3.23 Program clock reference

A transport stream is a multiplex of several TV programs and these may have originated from widely different locations. It is impractical to expect all the programs in a transport stream to be genlocked and so the stream is designed from the outset to allow unlocked programs. A decoder running from a transport stream has to genlock to the encoder and the transport stream has to have a mechanism to allow this to be done independently for each program. The synchronizing mechanism is called program clock reference (PCR).

Figure 3.46 shows how the PCR system works. The goal is to re-create at the decoder a 27 MHz clock synchronous with that at the encoder. The encoder clock drives a 48-bit counter that continuously counts up to the maximum value before overflowing and beginning again.

A transport stream multiplexer will periodically sample the counter and place the state of the count in an extended packet header as a PCR (see Figure 3.45). The demultiplexer selects only the PIDs of the required program, and it will extract the PCRs from the packets in which they were inserted.

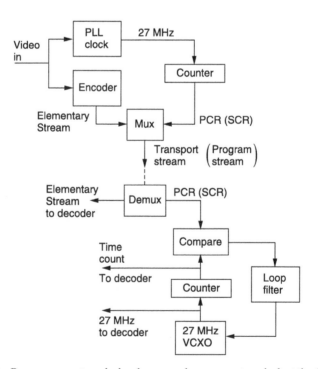

Figure 3.46 Program or system clock reference codes regenerate a clock at the decoder. See text for details.

The PCR codes are used to control a numerically locked loop (NLL). The NLL contains a 27 MHz VCXO (voltage controlled crystal oscillator), a variable-frequency oscillator based on a crystal having a relatively small frequency range.

The VCXO drives a 48-bit counter in the same way as in the encoder. The state of the counter is compared with the contents of the PCR and the difference is used to modify the VCXO frequency. When the loop reaches lock, the decoder counter would arrive at the same value as is contained in the PCR and no change in the VCXO would then occur. In practice the transport stream packets will suffer from transmission jitter and this will create phase noise in the loop. This is removed by loop filter so that the VCXO effectively averages a large number of phase errors.

A heavily damped loop will reject jitter well, but will take a long time to lock. Lockup time can be reduced when switching to a new program if the decoder counter is jammed to the value of the first PCR received in the new program. The loop filter may also have its time constants shortened during lockup. Once a synchronous 27 MHz clock is available at the decoder, this can be divided to provide the 90 kHz clock frequency that drives the time stamp mechanism.

The entire timebase stability of the decoder is no better than the stability of the clock derived from PCR. MPEG-2 sets standards for the maximum amount of jitter present in PCRs in a real transport stream.

Clearly, if the 27 MHz clock in the receiver is locked to one encoder it can only receive elementary streams encoded with that clock. If it is attempted to decode, for example, an audio stream generated from a different clock, the result will be periodic buffer overflows or underflows in the decoder. Thus MPEG defines a program in a manner that relates to timing. A program is a set of elementary streams encoded with the same master clock.

3.24 Transport stream multiplexing

A transport stream multiplexer is a complex device because of the number of functions it must perform. A fixed multiplexer will be considered first. In a fixed multiplexer, the bit rate of each of the programs must be specified so that the sum does not exceed the payload bit rate of the transport stream. The payload bit rate is the overall bit rate less the packet headers and program specific information (PSI) rate.

In practice, the programs will not be synchronous to one another, but the transport stream must produce a constant packet rate given by

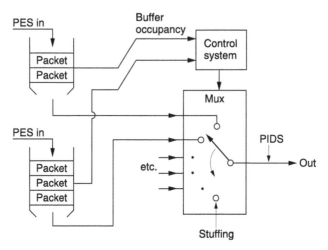

Figure 3.47   A transport stream multiplexer can handle several programs which are asynchronous to one another and to the transport stream clock. See text for details.

the bit rate divided by 188 bytes, the packet length. Figure 3.47 shows how this is handled. Each elementary stream entering the multiplexer passes through a buffer divided into payload-sized areas. Note that periodically the payload area is made smaller because of the requirement to insert PCR.

MPEG-2 decoders also have a quantity of buffer memory. The challenge to the multiplexer is to take packets from each program in such a way that neither its own buffers nor the buffers in any decoder overflow or underflow. This requirement is met by sending packets from all programs as evenly as possible rather than bunching together a lot of packets from one program. When the bit rates of the programs are different, the only way this can be handled is to use the buffer contents indicators. The more full a buffer is, the more likely it should be that a packet will be read from it. Thus a buffer content arbitrator can decide which program should have a packet allocated next.

If the sum of the input bit rates is correct, the buffers should all slowly empty because the overall input bit rate has to be less than the payload bit rate. This allows for the insertion of PSI. Whilst PATs and PMTs are being transmitted, the program buffers will fill up again. The multiplexer can also fill the buffers by sending more PCRs as this reduces the payload of each packet. In the event that the multiplexer has sent enough of everything but still can't fill a packet then it will send a null packet with a PID of 8191. Decoders will discard null packets and as they convey no useful data, the multiplexer buffers will all fill whilst null packets are being transmitted.

The use of null packets means that the bit rates of the elementary streams do not need to be synchronous with one another or with the transport stream bit rate. As each elementary stream can have its own PCR, it is not necessary for the different programs in a transport stream to be genlocked to one another; in fact they don't even need to have the same frame rate.

This approach allows the transport stream bit rate to be accurately defined and independent of the timing of the data carried. This is important because the transport stream bit rate determines the spectrum of the transmitter and this must not vary.

In a statistical multiplexer or statmux, the bit rate allocated to each program can vary dynamically. Figure 3.48 shows there must be a tight connection between the statmux and the associated compressors. Each compressor has a buffer memory which is emptied by a demand clock from the statmux. In a normal, fixed bit rate, coder the buffer content feeds back and controls the requantizer. In statmuxing this process is less severe and only takes place if the buffer is very close to full, because the degree of coding difficulty is also fed to the statmux.

The statmux contains an arbitrator that allocates more packets to the program with the greatest coding difficulty. Thus if a particular program encounters difficult material it will produce large prediction errors and begin to fill its output buffer. As the statmux has allocated more packets to that program, more data will be read out of that buffer, preventing overflow. Of course this is only possible if the other programs in the transport stream are handling typical video.

In the event that several programs encounter difficult material at once, clearly the buffer contents will rise and the coders will have to increase their compression factors.

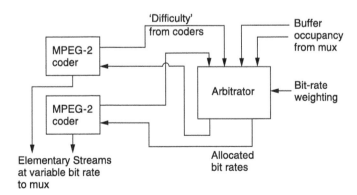

Figure 3.48 A statistical multiplexer contains an arbitrator which allocates bit rate to each program as a function of program difficulty.

Appendix 3.1 Calculation of Reed–Solomon generator polynomials

For a Reed–Solomon codeword over GF(2^3), there will be seven three-bit symbols. For location and correction of one symbol, there must be two redundant symbols P and Q, leaving A–E for data.

The following expressions must be true, where a is the primitive element of $x^3 \oplus x \oplus 1$ and $\oplus$ is XOR throughout:

$$A \oplus B \oplus C \oplus D \oplus E \oplus P \oplus Q = 0 \tag{1}$$

$$a^7 A \oplus a^6 B \oplus a^5 C \oplus a^4 D \oplus a^3 E \oplus a^2 P \oplus a Q = 0 \tag{2}$$

Dividing equation (2) by a:

$$a^6 A \oplus a^5 B \oplus a^4 C \oplus a^3 D \oplus a^2 E \oplus a P \oplus Q = 0$$

$$= A \oplus B \oplus C \oplus D \oplus E \oplus P \oplus Q$$

Cancelling Q, and collecting terms:

$$(a^6 \oplus 1)A \oplus (a^5 \oplus 1)B \oplus (a^4 \oplus 1)C \oplus (a^3 \oplus 1)D \oplus (a^2 \oplus 1)E$$

$$= (a \oplus 1)P$$

Using section 3.16 to calculate $(a^n + 1)$, e.g. $a^6 + 1 = 101 + 001 = 100 = a^2$:

$$a^2 A \oplus a^4 B \oplus a^5 C \oplus a D \oplus a^6 E = a^3 P$$

$$a^6 A \oplus a B \oplus a^2 C \oplus a^5 D \oplus a^3 E = P$$

Multiplying equation (1) by a^2 and equating to equation (2):

$$a^2 A \oplus a^2 B \oplus a^2 C \oplus a^2 D \oplus a^2 E \oplus a^2 P \oplus a^2 Q = 0$$

$$= a^7 A \oplus a^6 B \oplus a^5 C \oplus a^4 D \oplus a^3 E \oplus a^2 P \oplus a Q$$

Cancelling terms $a^2 P$ and collecting terms (remember $a^2 \oplus a^2 = 0$):

$$(a^7 \oplus a^2)A \oplus (a^6 \oplus a^2)B \oplus (a^5 \oplus a^2)C \oplus (a^4 \oplus a^2)D \oplus$$

$$(a^3 \oplus a^2)E = (a^2 \oplus a)Q$$

Adding powers according to section 3.16, e.g.

$$a^7 \oplus a^2 = 001 \oplus 100 = 101 = a^6:$$

$$a^6 A \oplus B \oplus a^3 C \oplus aD \oplus a^5 E = a^4 Q$$

$$a^2 A \oplus a^3 B \oplus a^6 C \oplus a^4 D \oplus aE = Q$$

References

1. Tang, D.T., Run-length-limited codes. IEEE International Symposium on Information Theory (1969)
2. AES Recommended practice for Digital Audio Engineering – Serial Multichannel Audio Digital Interface (MADI). *J. Audio Eng. Soc.*, **39**, No. 5, 371–377 (1991)
3. Shannon, C.E., A mathematical theory of communication. *Bell System Tech. J.*, **27**, 379 (1948)
4. Reed, I.S. and Solomon, G., Polynomial codes over certain finite fields. *J. Soc. Indust. Appl. Math.*, **8**, 300–304 (1960)
5. Berlekamp, E.R., *Algebraic Coding Theory*. New York: McGraw-Hill (1967). Reprint edition: Laguna Hills, CA: Aegean Park Press (1983)
6. Sugiyama, Y. *et al.*, An erasures and errors decoding algorithm for Goppa codes. *IEEE Trans. Inf. Theory*, **IT–22** (1976)
7. Peterson, W.W. and Weldon, E.J., *Error Correcting Codes*, 2nd edn, Cambridge MA: MIT Press (1972)

4

Dedicated audio interfaces

For the purposes of this book, digital audio interfaces will be divided into two types. This chapter is concerned with those that are dedicated point-to-point audio interfaces (e.g. AES/EBU), designed specifically to carry audio and little else. The next chapter deals with those that are protocols running over general purpose data networks or are computer interfaces that can also carry other data (e.g. USB).

Some interfaces have been internationally standardized whereas others are associated principally with one manufacturer. Some interfaces of the latter type have become widely used by other manufacturers, either because no alternative existed at the time or in order to provide compatibility between devices. It is important to distinguish *de facto* standards, which have arisen because of commercial predominance, from the standards formulated by independent international bodies. Proprietary interfaces may be the subject of licences or patents, although most agree that wide adoption is beneficial for all concerned. A further subdivision of interface types is also useful, and that is between interfaces carrying one or two channels of audio data and those carrying a large number of channels.

Naturally, since standards are published and widely available, these chapters should be viewed as commentaries upon or illuminations of the published documents, together with guidelines on their implementation and discussions of the problems involved when attempting to interconnect devices digitally. Although some details of standards will be given here, readers are encouraged to read the information contained herein in conjunction with the standards documents themselves (see the references at the end of each chapter), and to note any additions or modifications to the standards which may have arisen since this book was written. As this book is designed to aid understanding of standards in real situations it is not worded like a standards document and therefore does not use the official language of such documents. It is not a substitute for the standards themselves and any intending implementer would be wise not to rely solely on the text and diagrams contained herein.

4.1 Background to dedicated audio interfaces

When digital audio interfaces were first introduced it was assumed that digital audio signals would need to be carried between devices on connections similar to those used for analog signals. In other words, there would need to be individual point-to-point connections between each device, carrying little other than audio and using standard connectors and cables. The AES3 interface, as described below, is a good example of such an interface. It was intended to be used in as similar a way as possible to the method used for connecting pieces of analog audio equipment together. It used XLR connectors with relatively standard cables so that existing installations could be converted to digital applications. From a practical point of view, the only practical difference was that a single connection carried two channels instead of one. Such dedicated interfaces carry one or more channels of audio data, normally sample-locked to the transmitting device's sampling rate, and operate in a real-time 'streaming' fashion. They generally do not operate using an addressing structure, and so are normally used for connections between a single transmitting device and a single receiving device (hence 'point-to-point'). This method of interconnection is still in wide use today and is likely to continue to be so for some time, but the increasing ubiquity of high-speed data networks and computer-based audio systems is likely to have an increasing effect on the way audio is carried as time goes by (see the next chapter).

4.2 Background to internationally standardized interfaces

Within the audio field the Audio Engineering Society (AES) has been a lead body in determining digital interconnect standards. Although the AES is a professional society, and not a standards body as such, its recommendations first published in the AES3-1985 document[1] have formed the basis for many international standards documents concerning a two-channel digital audio interface. The Society has been instrumental in coordinating professional equipment manufacturers' views on interface standards although it has tended to ignore consumer applications to some extent, preferring to leave those to the IEC (see below). A consumer interface was initially developed by Sony and Philips, subsequently to be standardized by the IEC and EIAJ, and as a result there are many things in common between the professional and consumer implementations. Before setting out to describe the international standard two-channel interface it is important to give a summary of the history of the standard, since it will then be realized how difficult it is to call this interface by one definitive title.

Other organizations that based standards on AES3 recommendations were the American National Standards Institute (ANSI), the European Broadcasting Union (EBU), the International Radio Consultative Committee (CCIR) (now the ITU-R), the International Electrotechnical Commission (IEC), the Electronic Industries Association of Japan (EIAJ) and the British Standards Institute (BSI). Each of these organizations formulated a document describing a standard for a two-channel digital audio interface, and although these documents were all very similar there were often also either subtle or (in some cases) not-so-subtle differences between them. As time has gone by some of the most glaring anomalies have been addressed. A useful overview of the evolutionary process that resulted in each of these standards may be found in Finger[2]. The documents concerned were as follows: AES3-1985[1], ANSI S4.40-1985, EBU Tech. 3250-E[3] (1985), CCIR Rec. 647 (1986)[4], CCIR Rec. 647 (1990)[5], IEC 958 (1989)[6] (with subsequent annexes), EIAJ CP-340 (1987)[7], EIAJ CP-1201 (1992)[8], and BS 7239 (1989)[9].

As mentioned above, the roots of the consumer format interface were in a digital interface implemented by Sony and Philips for the CD system in 1984. This interface was modelled on the data format of AES3, but used different electrical characteristics (see section 4.3.4) and is often called the SPDIF (Sony–Philips Digital Interface). Although audio data was in the same format as AES3, there were significant differences in the format of non-audio data. In 1987 the EIAJ CP-340 standard subsequently combined professional (Type I) and consumer (Type II) versions of the interface within one document and included specifications for non-audio data which aimed to ensure compatibility between different consumer devices such as DAT and CD players. This interface was by no means identical to the original SPDIF and led to some differences between early CD players and later digital devices conforming to CP-340. CP-340 has now been renumbered and is called CP-1201.

Slightly later than CP-340 the IEC produced a document that eventually appeared in 1989 as IEC 958. The consumer version (with its subsequent annexes) was an extension of the SPDIF to allow for wider applications than just the CD, and the professional version was essentially the same as AES3. It also allowed for, but did not describe in detail, an optical connection. As will be seen later, interpreting this standard to the letter seemingly allowed the manufacturer to combine either consumer or professional *data* formats with either 'consumer' or 'professional' *electrical* interfaces. It did not originally state that a particular electrical interface had to be used in conjunction with a particular data format, although the situation was made clearer in the revised IEC 60958 standard[10]. Some confusion therefore existed in the industry over whether consumer devices could be interconnected with professional and vice versa, and the answer to this problem is by no means straightforward, as will be discussed.

Concerning key similarities and differences between the other documents listed above, one should note that EBU Tech. 3250-E is a professional standard, the only effective difference from AES3 being the insistence on the use of transformer coupling. Tech. 3250-E was revised in 1992 to define more aspects of the channel status bits, to allow a speech quality coordination channel in the auxiliary bits, to give an improved electrical specification and to specify which aspects of the interface should be implemented in standard broadcast equipment. CCIR Rec. 647 was a professional standard that did not insist on transformers, but was otherwise similar to the EBU standard. The 1990 revision contained some further definitions of certain non-audio bits, including use of the auxiliary bits for a low quality coordination channel (see section 4.5). CCIR Rec. 467 became ITU-R BS.647 (1992) when the CCIR was reborn as the ITU-R. BS 7239 was identical to IEC 958. ANSI S4.40 was identical to AES3. (The formal relationship between ANSI and AES standards has recently been broken.)

It may reasonably be concluded from the foregoing discussion that a device which claimed conformity to AES3, ANSI S4.40, EBU 3250 or CCIR 647 would have been be a professional device, but that one which claimed conformity to IEC 958, EIAJ CP-340, 1201 or BS 7239 could have been either consumer or professional. It was conventional in the latter case to specify Type I or II, for professional or consumer. (In modern IEC nomenclature, the consumer application is defined in 60958-3 and the professional application is in 60958-4.) Because it will be necessary to refer to differences between these standards in the following text, the cumbersome but necessary term 'standard two-channel interface' will be used wherever generalization is appropriate.

Some revisions of the original documents have taken place, the most important of which will be found in AES3-1992 (with revisions in 1997 and amendments up to 1999)[11] and IEC 60958 (superseding IEC 958 and covering general specifications, software delivery mode, consumer and professional applications in four parts). Because these two standards are the most comprehensive and form the primary focal points for international standards activity relating to two-channel interfaces, most of the following discussion will deal primarily with AES3 for professional interfaces and IEC 60958 for all other purposes. (The British Standard now follows the IEC standard directly in content and nomenclature, being denoted BS EN 60958 (2000).)

Owing mainly to the efforts of four UK companies, a further standard was devised to accommodate up to 56 audio channels. Originally called MADI (Multichannel Audio Digital Interface), it is based on the AES3 data format and has now been standardized as AES10-1991[12]. It also appears as an American Standard: ANSI S4.43-1991. This is only a professional interface and is covered further in section 4.11.

4.3   Standard two-channel interface – principles

Common to all the international standards for a two-channel interface is the data format of the subframe containing samples of audio data for each channel. There are two principal electrical approaches used for the standard two-channel interface: one is unbalanced and uses relatively low voltages, the other is balanced and uses higher voltages. AES3-ID-2001 also describes an unbalanced coaxial link for use over distances beyond 100 m (see section 4.3.6).

4.3.1   Data format

The interface is serial and self-clocking. That is to say that two channels of audio data are carried in a multiplexed fashion over the same communications channel, and the data is combined with a clock signal in such a way that the clock may be extracted at the receiver and used to synchronize reception. As shown in Figure 4.1, one frame of data is divided into two subframes, handling channels 1 and 2 respectively. Channels 1 and 2 may be independent mono signals or they may be the left and right channels of a stereo pair, and they are separately identified by the preamble that takes up the first four clock periods of each subframe. Samples of channels 1 and 2 are transmitted alternately and in real time, such that two subframes are transmitted within the time period of one audio sample – thus the data rate of the interface depends on the prevailing audio sampling rate.

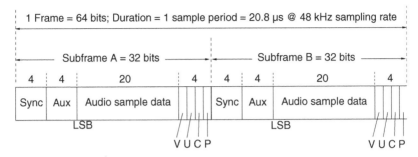

Figure 4.1 Format of the standard two-channel interface frame.

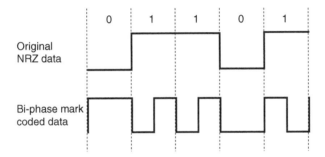

Figure 4.2 An example of the bi-phase mark channel code.

The subframe format consists of a sync preamble, four auxiliary bits (which may be used for additional audio resolution), 20 audio sample bits in linear two's complement form, a validity bit (V), a user bit (U), a channel status bit (C) and a parity bit (P). The audio data is transmitted least significant bit first, and any unused LSBs are set to zero; thus the MSB of the audio sample, whatever the resolution, is always in the MSB position. The remaining non-audio bits are discussed in later sections.

The data is combined with a clock signal of twice the bit rate using a simple coding scheme known as *bi-phase mark*, in which a transition is caused to occur at the boundary of each bit cell (see Figure 4.2). An additional transition is also introduced in the middle of any bit cell that is set to binary state '1'. Such a scheme eliminates almost all DC content from the signal, making it possible to use transformer coupling if necessary and allowing for phase inversion of the data signal. (It is only the transition that matters, not the direction of the transition.) This channel code is the same as that used for SMPTE/EBU timecode.

As shown in Figure 4.3, there are three possible subframe preambles in time slots 1 to 4 which violate the rules of the modulation scheme in order to provide a clearly recognizable sync point when the data is decoded.

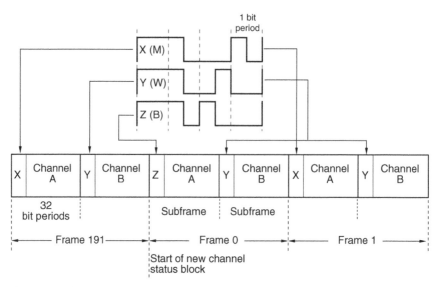

Figure 4.3   Three different preambles (X, Y and Z) are used to synchronize a receiver at the starts of subframes.

These preambles cannot be confused with the data portion of the subframe. In AES3 these are called 'X', 'Y' and 'Z' preambles, but in IEC 60958 they are primarily labelled 'M', 'W' and 'B'. As the diagram shows, X and Y preambles identify subframes of channels 1 and 2 respectively, whereas the Z preamble occurs once every 192 frames in place of the X preamble in order to mark the beginning of a new channel status block (see section 4.8). Since the parity bit which ends the previous subframe is 'even parity', the transition at the start of each preamble will always be in the same (positive) direction, but a phase inverted preamble must still be decoded properly.

The parity bit is set such that the number of ones in the subframe, excluding the preamble, is even, and thus it may be used to detect single bit errors but not correct them. Such a parity scheme cannot detect an even number of errors in the subframe, since parity would appear to be correct in this case. As discussed in section 6.9 there are more effective ways of detecting poor links than using the parity bit.

4.3.2   Audio resolution

In normal operation only the 20-bit chunk of the subframe is used for audio data.

This is adequate for most professional and consumer purposes but the standard allows for the four auxiliary bits to be replaced by additional audio LSBs if necessary, taking the maximum resolution up to 24 bits. AES3-1992 provides a facility within the channel status data for signalling the actual number of audio bits used in the transmitted data, such that receiving equipment may adjust to decode them appropriately. This will be of considerable importance in ensuring optimum transfer of audio quality between devices of different resolutions during post-production, as discussed in section 6.8.

In consumer formats, the category code that describes the source device (see section 4.8.6) may also imply a fixed audio word length, because certain categories only operate at a particular resolution. The Compact Disc, for example, always uses a 16-bit word length.

4.3.3 Balanced electrical interface

All the standards referring to a professional or 'broadcast use' interface specify a balanced electrical interface conforming to CCITT Rec. V.11[13]. There are distinct similarities between this and the RS-422A standard[14] but they are not identical, although RS-422 drivers and receivers are used in many cases. Figure 4.4 shows a circuit designed for better isolation and electrical balance than the basic CCITT specification, as suggested in AES3-1992. Although transformers are not a mandatory feature of all the standards, they are advisable because they provide true electrical isolation between devices and help to reduce electromagnetic interference problems. (Manufacturers often connect an RS-422 driver directly between the two legs of the source, which makes it balanced but not floating. Alternatively an RS-485 driver is used, which is a tri-state version of RS-422 giving a typical output voltage of 4 V ± 5%, going to a high impedance state when turned off.) The standards specify that the connector to be used is the conventional audio three-pin XLR (IEC 268-12), using pin 1 as the shield and pins 2 and 3 as the balanced data signal. Polarity is not really

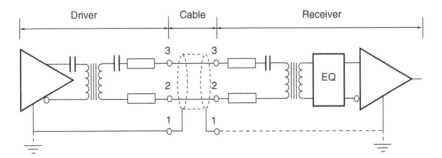

Figure 4.4   Recommended electrical circuit for use with the standard two-channel interface.

important, since the channel code is designed to allow phase inversion, although the convention is that pin 2 is '+' and pin 3 is '−'.

Although the original AES3 standard allowed for up to four receivers to be connected to one transmitter, this is now regarded as inadvisable due to the impedance mismatch which arises. Originally the standard called for the output impedance of the transmitter to be 110 ohms ± 20% over the range 0.1 to 6 MHz, and for that of receivers to be 250 ohms, but this has been changed in AES3-1992 so that the receiver's impedance should now be the same as that of the transmitter and the transmission line. Amendment 3 (1999) modifies the specification to accommodate the increasingly common use of higher frame rates than originally envisaged, as a result of the use of high sampling frequencies such as 96 kHz. The standard now specifies that impedance should be maintained within the defined limits between 100 kHz and 128 times the maximum frame rate. Only one receiver should be connected across each line and distribution amplifiers should be used for feeding large numbers of receivers from a single source. The cable's characteristic impedance, originally specified as between 90 and 120 ohms, is now specified as 110 ohms as well. It should be a balanced, screened pair, and although standard audio cables are often used successfully it is worthwhile considering cable with better controlled characteristics for large installations and long distances, in order to improve the reliability and integrity of the link (see section 6.7.2). This is especially true when using sampling frequencies above 48 kHz where the selection of cables and maximum lengths will become increasingly critical.

There is a difference in driver voltage levels between the original and later versions of the standard. AES3-1985 and all the related standards specified a peak-to-peak amplitude of between 3 and 10 volts when measured across a 110 ohm resistor without the connecting cable present. The 1992 revision changed it to be between 2 and 7 volts in order to conform more closely to the specifications of the RS-422 driver chips used in many systems. (RS-422A in fact specifies that receiver inputs should not be damaged by voltages of less than 12 volts.)

At the receiving end, the standards all indicate that correct decoding of the data should be possible provided that the eye pattern (see section 3.2.4) of the received data is no worse than shown in Figure 4.5. This suggests a minimum peak-to-peak amplitude of 200 mV and allows for the toleration of a certain amount of jitter in the time domain. Without equalization the balanced interface should be capable of error-free communication over distances of at least 100 m at 48 kHz sampling frequency, and often further. This depends to some extent on the type of cable, the electromagnetic environment, the integrity of the transmission line, the frame

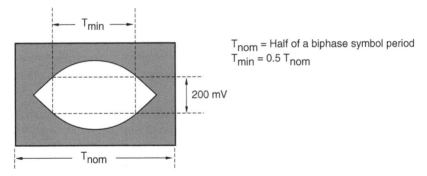

Figure 4.5 The minimum eye pattern acceptable for correct decoding of standard two-channel data.

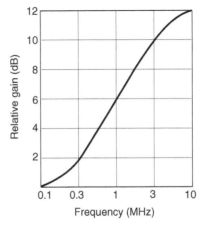

Figure 4.6 EQ characteristic recommended by the AES to improve reception in the case of long lines (basic sampling rate).

rate and the quality of the data recovery in the receiver. One should expect maximum cable lengths to be shorter at high frame rates, all other factors being equal. Receivers vary quite widely in respect of their ability to lock to an unstable data signal which has suffered distortion over the link, and an interconnect which works badly with one receiver may be satisfactory with another. Devices are available which will give some idea of the quality of the received data signal, in order that the user may tell how close the link is to failure (see section 6.9).

It is possible to equalize the signal at the receiver in order to compensate for high-frequency losses over long links and the standards suggest the curve shown in Figure 4.6 for use at the 48 kHz sampling frequency. It has been suggested[15], though, that as cable lengths increase the loss characteristic approaches a second order curve before problems occur, and that therefore a second order equalization characteristic is often more effective.

4.3.4 Unbalanced electrical interface

The unbalanced interface described in this section is commonly found on consumer and semi-professional equipment and has become widely used as a stereo interface on computer sound cards, probably because of the compact size of the connector. The unbalanced electrical interface specified originally in IEC 958 and EIAJ CP-340/1201 is not a feature of professional standards such as AES3. IEC 958 did not originally state explicitly that the unbalanced interface was intended for consumer use – it simply called it 'unbalanced line (two-wire transmission)' – but operational convention and the origin of the SPDIF interface on which it was based established that the unbalanced two-wire interface, terminating in RCA phono connectors, was for consumer applications. Interestingly, EIAJ CP-340 took the step of noting that the unbalanced interface and the optical fibre interface applied only to Type II transmissions (consumer), although it did not say anything about the balanced interface being only for professional purposes. These confusions are resolved in IEC 60958 which clearly indicates the use of the unbalanced or optical interfaces for consumer applications in Part 3.

The unbalanced interface is shown in Figure 4.7. IEC 60958 (1999) specifies a source impedance of 75 ohms ± 20% for this interface, between 0.1 and 6 MHz, and a termination impedance of 75 ohms ± 5%. Like AES3, it is being revised to account for higher sampling frequencies and so will in future state an upper limit of 128 times the maximum frame rate. It specifies a characteristic cable impedance of 75 ohms ± 35%. The cable is normally a standard audio coaxial cable and this interface is typically used for interconnecting consumer equipment over the sorts of distances involved in hi-fi systems. It does not specify a maximum length over which communication may be expected to be successful but it does give an eye pattern limit for correct decoding and specifies a minimum peak-to-peak input voltage at the receiver of 200 mV (the minimum eye pattern for correct decoding is essentially the same as AES3). A significant difference between this interface and the balanced interface is that the source signal amplitude should be only 0.5 V ± 20%, peak-to-peak, which is much lower than the balanced interface. It should be noted, though, that video-type 75 ohm coaxial cable exhibits very low losses below about 10 MHz and so one might expect to be able to cover significant distances without the signal level falling below the minimum specified.

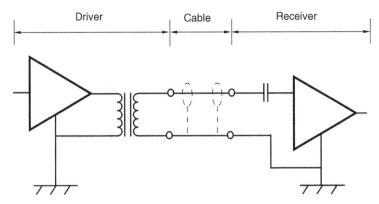

Figure 4.7 The consumer electrical interface (transformer and capacitor are optional but may improve the electrical characteristics of the interface).

It used to be said by some that because the unbalanced interface was a coaxial transmission line with well-controlled impedances it formed a better link than the balanced interface. This was always offset by the advantages of a balanced line in rejecting interference and the higher voltages used in the balanced interface. Now that the balanced interface specifies source and termination impedances to be the same, requires point-to-point connection, and recommends 110 ohm cable (rather than anything between 90 and 120 ohms), the balanced interface has the benefits of a good transmission line as well as its other advantages.

4.3.5 Optical interface

An optical interface was introduced as a possibility in IEC 958 but was left 'under consideration'. Surprisingly, perhaps, this still seems to be the case in 60958. It was specified more explicitly in EIAJ CP-340 (or CP-1201) as applying only to Type II data and consisting of a transmitter with a wavelength of 660 nm ± 30 nm and a power of between −15 and −21 dBm. Receivers should still correctly interpret the data when the optical input power is −27 dBm. The connector indicated conforms to the specification laid out in EIAJ RCZ-6901.

Typically the optical interface is found in consumer equipment such as DAT recorders, CD players, computer sound cards, stand-alone convertors, and amplifiers with built-in D/A convertors. It usually makes use of an LED transmitter (see section 1.7) and a fibre optic cable, connected to a photodetector in the receiver. The 'TOSLink' style of fibre optic interface is popular in consumer equipment, and is driven from a TTL level (0–5 volt) unbalanced source, with a data format identical to that used with the electrical interface. The advantages of optical links in rejecting interference have already been stated in section 1.7 but there are also dangers in using cheap optical interfaces. Their limited bandwidth and high dispersion may actually result in a poorer transmission channel than a normal electrical interface, resulting in a high degree of timing instability in the positions of data transitions (see Chapter 6). For a comprehensive introduction to fibre optics in audio the reader is referred to Ajemian and Grundy[16].

4.3.6 Coaxial interface

A coaxial method of transmission for the professional AES3 interface, described in AES-3ID[17], makes use of 75 ohm video-style coaxial cable to carry digital audio signals over distances up to around 1000 m. A similar but not identical description of this is to be found in SMPTE 276M[18]. A signal level similar to that of a video signal (1 volt) is used, although the signal is not formatted to look like a video waveform (it still uses basically the same bi-phase mark channel code as AES3). Easy conversion is possible between the balanced form of AES3 and this coaxial form and a number of manufacturers make balanced-to-coax adaptors (although the voltage level is much lower after transformation from 1 volt/75 ohms back to 110 ohms than might be expected from a standard AES3 balanced output stage). Some simple conversion networks are illustrated in the information document.

The advantages of this interface include the ease of distribution of audio within a television studio environment, using video distribution amplifiers and cabling, and improved electromagnetic radiation characteristics when compared with the balanced twisted pair, as discussed in Rorden and Graham[19]. Tests of such an interface have been successful, showing in one test that equalized video lines could carry an AES-format digital audio signal over a distance of more than 800 miles (1300 km) without any noticeable corruption.

4.3.7 Multipin connector

A multipin connector version of AES3 is described in the information document AES2-ID[20] for use in circumstances in which there is not sufficient space for multiple XLR connectors. Such a connector allows multiple channel interfacing without going to the lengths of implementing the MADI standard (see section 4.11) and could be a lower cost solution than MADI in cases of a smaller number of channels than 56. The described configuration carries 16 channels of audio data on a single 50 pin D-type connector.

4.4 Sampling rate related to data rate

The standard two-channel interface was originally designed to accommodate digital audio signals with sampling rates between 32 and 48 kHz, with a margin of ±12.5% to allow for varispeed operations. Since the interface carries audio data in real time, normally transferring two audio samples (channel 1 and channel 2) in the time of one sampling period, the data rate of the interface depends on the audio sampling rate. It is normally 64 times the sampling rate, since there are 64 bits in a frame (= two subframes). At a sampling rate of 48 kHz the data rate is 64 times 48 000, which is 3.072 Mb/s, whereas at 32 kHz it is only 2.048 Mb/s. If the source is varispeeded by a certain percentage then the data rate will change by the same percentage, and although it can usually be tracked by a receiver this presents problems in a system where all devices must be locked to a common, fixed sampling frequency reference (see Chapter 6), since the receiver may not change its sampling rate to follow a varispeeded source.

In recent years there has been a demand for interfaces capable of handling audio at increased sampling frequencies up to 192 kHz. For this reason a situation can arise in which one AES3 interface is used in a single-channel-double-sampling-frequency mode. Here the two subframes within a single AES frame carry successive samples of the same audio channel, making the audio sampling frequency twice the AES frame rate. The sampling frequency indicated in byte 0 of channel status (see below) remains as if the interface was operating at normal sampling frequency, because this is the frame rate of the interface. For example, in this mode a single interface could carry a single channel of audio at 96 kHz sampling frequency whilst continuing to operate at the same data transmission rate as a stereo interface running at 48 kHz. This is a useful alternative to doubling the overall interface transmission rate to accommodate the higher audio sampling frequency (giving rise to the need for better and possibly shorter cables). These modes are indicated in channel status byte 1, described below.

4.5 Auxiliary data in the standard two-channel interface

As stated earlier, the four bits of auxiliary data in each subframe may be used for additional LSBs of audio resolution, if more than 20 bits of audio data are needed per sample. Alternatively the auxiliary data may be used to carry information associated with the audio channel. In many items of equipment manufactured to date they remain unused.

It was proposed to the CCIR in 1987 that the aux bits would prove useful for a good voice quality channel which could be used for coordination (talkback) purposes in broadcasting[21]. Typically in a radio broadcast studio the programme source (say a studio) sends a stereo programme to a destination (say a continuity suite) along with a good voice quality link for coordination purposes (see Figure 4.8). A feed of cue programme (normally mono), together with a coordination channel and perhaps additional data, is returned from the destination. It was proposed that in digital studio environments all of these signals could be carried over a single standard two-channel interface in each direction by sampling the coordination voice channel at exactly one-third of the main audio's sampling frequency and coding it linearly at 12 bits per sample, resulting in a data rate exactly one-fifth that of the main audio channel. (Main audio channel @ 48 kHz, 20 bits; Data rate = 960 000 bits per second; Coordination channel @ 16 kHz, 12 bits; Data rate = 192 000 bits per second.) At such a data rate, a main sampling rate of 48 kHz would allow for a coordination channel bandwidth of about 7 kHz.

Capacity exists in the two-channel interface for two 12-bit coordination samples, one in channel 1's subframe ('A' coordination) and one in channel 2's ('B' coordination). They are inserted four bits at a time, as shown in Figure 4.9, with the four LSBs of the 'A' signal going into the first aux word in the block (designated by the Z preamble), followed by the four LSBs of the 'B' signal in the next subframe, and so on for three frames, whereon all 12 bits will have been transmitted for each signal. The process then starts over again. The Z preamble thus acts as a sync point for the coordination channels, and the sampling frequencies of the coordination channels are locked to that of the main audio.

This arrangement proves very satisfactory because the correct talkback always accompanies a programme signal, the cue programme may be in stereo, and only two hard links are necessary. The user bit channel of the interface can be used to carry any additional control data. In the 1990 revision of CCIR Rec. 647 this modification was taken on board, and

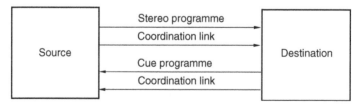

Figure 4.8 In broadcasting a coordination link often accompanies the main programme, and cue programme is fed back to the source, also with coordination.

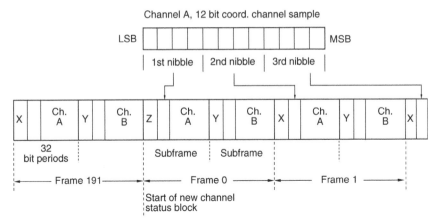

Figure 4.9 The coordination signal is of lower bit rate to the main audio and thus may be inserted in the auxiliary nibble of the interface subframe, taking three subframes per coordination sample.

forms part of that standard. It was also adopted by the EBU in the 1992 revision of Tech. 3250-E, and appears as an annex to AES3-1992 (for information purposes only). In order to indicate that the aux bits are being used for this purpose, byte 2 of the channel status word (see section 4.8) is used. The first three bits of this byte are set to '010' when a channel's aux bits carry a coordination signal of this type.

4.6 The validity (V) bit

The application and value of the validity bit in each subframe were debated widely during the formulation of standards. Originally the V bit was designed to indicate whether the audio sample in that subframe was 'valid' or 'reliable', 'secure and error free' – in other words, to show either whether it contained valid audio (rather than something else, or nothing), or if it was in error. It was set to '0' if the sample was reliable, and '1' if unreliable (so really it is an 'invalidity' flag). Since it is only a single bit, there is no opportunity for signalling the extent or severity of the error. What has never been clear is what devices should do in the case of an invalid sample and this is largely left up to the manufacturer. The most common use for the V bit is to signal errors that occurred in the transmitting device, such as when an uncorrectable error is encountered when replaying a recording, for example. But not all devices treat this in the same way, since some signal any offtape CRC error whether it was corrected or not, whereas others only set the V bit if the error was uncorrectable, resulting in interpolation or even muting in the convertors of most systems (this seems the more appropriate solution). But this is not the only use. Interactive CD (CD-I) players, for example, use the V bit to indicate that the audio data part of the subframe has been replaced by non-audio data. This is because otherwise there would be a potential delay of up to 1 block (192 frames) before the receiving device realized that the channel use had changed from audio to non-audio (this is normally signalled in channel status that is only updated once per block).

In AES3-1992 the description of this bit was modified and now indicates whether the audio information in the subframe is 'suitable for conversion to an analog signal', which may or may not amount to the same thing as before. (A binary '1' still represents the error state.) For example, the V bit may be set if an uncorrectable error arises on a DAT machine which would normally result in error concealment by interpolation rather than muting at the output. Any subsequent device receiving audio data from this machine over the digital interface would see the V bit set true and, if interpreting it literally, would assume that the audio was unsuitable for conversion and mute its output, yet the user might still want to hear it, assuming that the interpolation sounds better than a mute! Receivers vary in this respect, and there are some that always mute on seeing the V bit set true. A further and potentially more serious problem, highlighted by Finger[2], is that a recording device will usually store incoming audio data but has no means of storing the validity flag. In such a case the replayed audio would then be transmitted with no indication of invalidity, even if recorded errors existed.

Another problem arises in devices which simply process the data signal, such as the sample rate convertors or interface processors described in Chapter 6. Should such devices take any action in the case of invalid samples, or should they simply pass the data through untouched? One such device takes the approach of carrying through the V bit state and holding the last sample value in the case of an error, but the solution is less clear when two digital signals are to be mixed together, one which is erroneous and the other not. In such a case it is difficult to decide whether the single mixed data stream should be valid or invalid and there are no clear guidelines on the matter (indeed there is no 'catch-all' solution to such a problem).

In truth, the most appropriate actions to and uses of the V bit are application and product dependent. AES2-ID discusses some of these issues and concludes that it is largely the responsibility of the manufacturer to determine the most appropriate treatment of the V bit in accordance with its general definition. Since it is now recommended that all products implementing AES3 are provided with an implementation table, notes on the treatment of the V bit should be incorporated in a similar place in the relevant manual. It is not recommended that the V bit be used as a permanent alternative to the 'audio/non-audio' bit in the channel status block, as an indicator of ongoing non-audio data within the subframe, but this could be a temporary solution until the start of the next channel status block as noted above. IEC 60958 describes this as a good temporary use of the V bit for consumer applications involving non-audio data.

4.7 The user (U) channel

The U bit of each subframe has a multiplicity of uses, many of which have remained hidden from the user of commercial equipment, such as the carrying of text, subcode, and other non-audio data. It is most widely used in consumer equipment and there is now a rather complicated AES standard for its use in professional applications (AES18, see below). There is also a Philips method for inserting data into the user channel, called ITTS (Interactive Text Transmission System), on which the CD system relies for the transferring of subcode and other non-audio data over the consumer interface. The U bit is only a single bit in each subframe, potentially allowing a user channel to accompany each audio channel, and its definition in the various standards is normally 'for any other information'. The user bits are not normally aggregated over the same block length as channel status data (192 subframes), although they may be, but are often aggregated over different block lengths depending on the application, or may simply be used as individual flags. Many devices, especially professional ones, do not use them at all, although this may change in the future.

In the following sections a number of the most common applications for user data are outlined, although the standards do not really prohibit users or manufacturers using this capacity for alternative purposes. AES3-1992 signals the use of the user bits in byte 1, bits 4–7 of channel status as shown in Table 4.1. IEC 60958 recommends a common format for

Table 4.1 Indication of user bits format in channel status byte 1

Bits 4–7	User bits format
0000	Default, no user information
0001	192-bit block structure
0010	AES18
0011	User defined

the application of user bits, suggesting that the user bits in each subframe should be combined to make a single user bitstream for each interface.

(It may be noted that the terminology used to describe the bit number of a message in the user channel can be confusing, depending on whether the message is considered as running 'MSB to LSB' or vice versa. We shall refer to the first transmitted bit of a message byte as 'bit 0', but some documents refer to this as 'bit 7'.)

4.7.1 HDLC packet scheme (AES18-1992)

Unlike channel status data, user data may consist of a wide variety of different message types, and the AES working group on 'labels' decided that the best approach to the problem for professional users was to allow the user channel to be handled in a 'free format', such that its maximum capacity of 48 Kbit/s could be shared between applications, with user data multiplexed into 'packets' of information which would share the interface. The history of this goes back to 1986, and a proposal by Roger Lagadec of Sony suggested that user data was very different to channel status data and would not suffer the same block structure, requiring a more flexible approach in which some messages could be sent once with minimal delay, whereas others might be repeated at regular time intervals, and yet others might have to be time-specific. It required that the data rate in the user channel be independent of the audio sampling rate, whereas the actual rate of user bits depends on the interface frame rate and thus on the sampling rate.

It is not necessary to document the whole history here, except to say that the direction of the work was influenced considerably by proposals from TDF (Télédiffusion de France) and others, well documented by Alain Komly[22], suggesting the use of an asynchronous frame format already well established in the telecommunications and computer industries called HDLC (High-level Data Link Control). This is an internationally standardized way of transferring data at a bit-oriented level around networks (ISO 3309-2)[23], and there are a number of commercial chips available which do the job of inserting data into the correct packet structure. The working group finally recommended a structure for carrying user data in AES18-1992[24], and a useful commentary on this may be found in Nunn[25]. The AES18 standard was revised in 1996[26] to include recommendations for coding the data carried over the user channel. If this particular way of treating the user bits is implemented then it is indicated in byte 1 of the channel status information, bits 4–7, as shown in Table 4.1.

Although this is a flexible and versatile way of treating the user bit channel it is possibly overcomplicated for some applications, leaving it up to the user to build his or her own applications around the protocol. It treats the channel rather like a transport stream on an asynchronous computer network and is probably most well suited to large broadcasting installations and systems, although it may quickly be overtaken by protocols that use standard high speed computer networks for both audio and data communications. This approach is not part of the consumer format.

Among the key features of this standard are that the data rate of the user channel can be kept constant over a defined range of sampling rates (but only between 42 and 54 kHz in AES18), that a precise timing relationship can be maintained between audio and user data, that time-critical data may be transmitted within a specified and guaranteed period, and that the channel may be used simultaneously by a number of users. User data to be transmitted is formed into packets which are preceded by a header containing the address of the destination, and the packet is then inserted into the user data stream as soon as there is room. In order to ensure that the user data rate remains constant down to an audio sampling rate of 42 kHz (which is 48 kHz minus 12.5%) extra packing bits are added at the end of each block of packets which can be disposed of as the sampling frequency is lowered. At audio sampling rates below 42 kHz the data rate will be lower, and thus some information would be lost if 48 kHz data were to be sample rate converted to, say, 44.1 or 32 kHz, but it is expected that some form of data management would be implemented to ensure that important data gets the highest priority in these circumstances.

Data is formatted at a number of levels before being transmitted over the interface, starting at the highest level – the 'application level' – and ending at the lowest level – the 'physical level' – at which the data is actually inserted bit by bit into the audio interface subframe structure. It is not intended to cover the process by which this is achieved here, since this would constitute needless repetition of available documentation. What is important is some commentary on the handling of different types of message, particularly time-specific messages, and on the insertion of additional messages at later points in the interface chain.

AES18 allows for the handling of time-specific messages by formatting the user data packets into blocks, normally of fixed but definable length, and repeating these at a user-definable rate which can be set to correspond to time intervals pertinent in the application concerned. An optional 'system packet' may also be transmitted at block intervals which may contain timecode data among other things, and sets priorities for different types of message which may have more or less urgency. It recommends some useful repetition rates of blocks, which correspond to the

Table 4.2 Some useful repetition rates of blocks

Blocks per second	Duration (ms)	Application
24	41.67	Film
25	40	PAL, SECAM video or 50 frame per second (fps) HDTV
29.97	33.37	NTSC video
30	33.33	60 fps HDTV
33.33	30	DAT

timing intervals of frames in audio and video applications, as shown in Table 4.2. In some applications variable block lengths may be necessary, such as when using 48 kHz audio with NTSC video (which runs at 29.97 fps) where there is not an integer number of audio samples per video frame.

In order to allow for the insertion of messages of varying importance at different points in the system, the standard sets down comprehensive rules governing the way in which messages should be prioritized. The maximum delay involved in inserting a packet of data depends on its priority (from 0 to 3), and the block length involved. The highest priority packet (level 3) may be inserted once per block, and as the priority is decreased the packets are inserted only once per so many blocks. Since the shortest practical block length is 10 ms, this is the minimum delay one might anticipate.

The original version of the standard defined packet structures and transport stream protocols, but said little or nothing about the format or structure of messages. AES18-1996 describes a means of addressing for messages that defines the application area and purpose of the message. Examples include messages about programme description, engineering notebook and switching information. Collaboration is claimed with the EBU regarding the format and structure of such messages.

4.7.2 Consumer applications of the user bit

IEC 60958 is more specific about recommended protocol for the user bit-stream than the original IEC 958, probably because the range of uses of the interface has grown greatly in the intervening period. In essence it suggests that the user bits for the two subframes should be combined to make a single data stream for the interface concerned. The basis for the recommended structure lies in the Compact Disc application of the user bits to transfer Q–W-channel subcode data, but it has been generalized to other product categories.

The relevant bits should be formed into information units (IUs) of eight bits, starting with a binary 1 and followed by seven information bits. Probably because of the historical link with CD subcode the eight bits of an IU are called the P–W bits, although the P bit does not bear relationship to the P-channel subcode data on CD. IUs are typically separated by four '0' bits, but can be separated by between none and eight. More than eight '0' bits in a row signifies the start of a new message. An example is shown in the next section, relating to the CD.

Three classes of equipment are indicated, each with a different role in relation to user bits, being essentially those that originally generate user bits (Class I), those that pass them through or are 'transparent' to user bits (Class II), and mixed-mode equipment such as may combine signals or process them (Class III). In the case of Class II equipment that delays the audio signal it is recommended that the user bits are similarly delayed in order to preserve time alignment.

Because of the somewhat disorganized history of user bit application in consumer products, different categories of products (indicated by the category code in channel status) have their own message structures. However, there is now a general structure that is recommended for new products that can start from scratch in their implementation. In the general structure a message may be made up of a minimum of three and a maximum of 129 IUs, although messages of 96 IUs are reserved specifically for certain classes of laser products. The first three IUs have a specific function, as shown in Figure 4.10, indicating the type of message, the number of remaining IUs in the message (after the first three) and the category code of the originating equipment. Message data is then carried in successive IUs. Because of the complexity and number of possibilities for user bit messaging in consumer products, the reader is referred to the standard and its annexes for further details, although most of the annexes show relatively little in this respect.

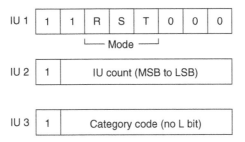

Figure 4.10 Format of the first three IUs of the consumer user bitstream.

4.7.3 Applications of the user bit in Compact Disc and MiniDisc systems

The following is given as a common example of the application of user bits in two consumer product categories. Compact Disc and MiniDisc players having consumer format digital outputs normally place subcode data in the user bits. This is in addition to the control bits of the Q-channel subcode from the disc which are transmitted within channel status (see section 4.8.9) and formed the only full specification for user bits implementation in the original IEC 958 document, as Annex A.

In this application the user bits for the left and right subframes are treated as one channel, and the Q to W subcode data is multiplexed between them (the P flag is not transmitted since it only represents positioning information for the transport). The subcode data block is built up over 1176 samples, formed into sync blocks of 12 samples each (making 98 subcode symbols, which include two symbols for block sync). There are eight subcode bits in each of these sync blocks (P–W), but only seven of them are transmitted (Q–W) over the interface. Because the subcode data rate is lower than the user bit rate, zeros are used as packing between the groups of subcode bits. The number of zeros is variable, principally to allow for variable speed replay to ±25%. As shown in Figure 4.11, the subcode block begins with a minimum of 16 zeros, followed by a start bit (a binary '1' which some documents call 'P' although this might be

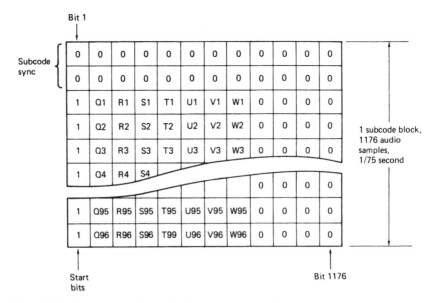

Figure 4.11 An example of user bits formatting in the CD system.

misleading since it is always a '1' and thus contains no additional information). There then follow seven subcode bits (Q1 to W1), after which there may be up to eight zeros before the next start bit and the next seven bits of subcode data. Only four packing zeros are shown in this diagram. This pattern is repeated 98 times, after which a new intermessage sync pattern of at least 16 zeros is expected.

The Q data in the subcode stream can be used to identify track starts and ends, among other things (see the full CD specification in IEC 908), so it is useful when transferring CDs to DAT or vice versa (for professional purposes, of course), or from a CD player to a CD recorder, since the audio data and the track IDs are duplicated together and the copy is a true clone of the original. Between CD machines there is usually little problem in copying subcode data, since the two machines are of the same format, but between CD and DAT a special processor unit is normally required to convert DAT track IDs to CD track IDs or vice versa and there are occasional discrepancies. Since the P flag is not transferred over the interface the copy may only rely on Q subcode information and there is usually a gap between the start of the P flag on the CD and the Q subcode track number increment. Some CD players increment the track number on their own displays at the start of the P flag and then count down to the true track start using the Q data, whereas a copy of such a recording would only increment the track number at the true track start. There is also occasionally a small delay in the assertion of the track start flag on DAT recordings, due to the automatic start ID facility used in many machines which writes a new start ID when the audio level rises above a certain point, which may sometimes be compensated for in the transfer.

4.7.4 Applications of the user bit in DAT systems

As with the CD/MD, the consumer interface on DAT machines also carries some additional information in the user bits. The first edition of IEC 958 suggested that subcode data would be carried in the four auxiliary bits rather than the user bits but IEC 60958 now shows subcode in the user bits, with nothing in the aux bits. Considering the subcode information which could be sent in the user bits the actual implementation is incredibly crude, as it simply indicates the presence or lack of start and skip (shortening) IDs on the tape. This approach was in fact inherent in the DAT design standard right from the start[27].

As shown in Figure 4.12, sync, start ID and skip ID are transmitted over the interface with relation to the DAT frame rate of 33.33 frames per second. As with CD, the user bits of the left and right channel subframes are

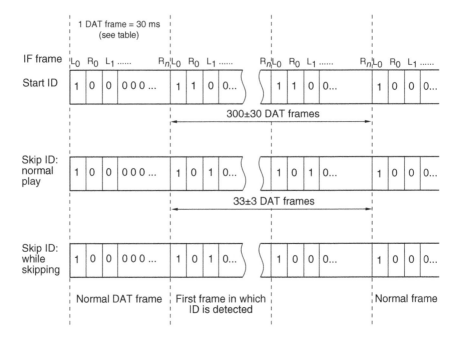

L_0 user bit is SYNC and goes true every DAT frame (every $n+1$ IF frames)

R_0 user bit is Start ID and is true for 300 ± 30 frames (i.e. for duration of Start ID on tape)

L_1 user bit is Skip ID and is true for 33 ± 3 frames when not skipping (just detecting a Skip ID on tape), but is true only for 1 frame at the start of an actual skip (in skip play)

Sampling rate	Digital IF frames per DAT frame	n
32 kHz	960	959
32 kHz (LP)	1920	1919
44.1 kHz	1323	1322
48 kHz	1440	1439

Figure 4.12 Signalling of DAT start and skip IDs in user bits (user bits only shown).

considered together and (differently from the 'general' format described above) each 'message' consists of only one IU with only two active bits (Q and R). The sync ID (P bit) is transmitted once per frame by setting the user bit of the first left channel sample (L_0) of that frame true – this simply indicates where the frame begins and could be used for crude synchronization of two machines. (In other words, the user bit of the interface subframe corresponding to the first sample of the DAT frame is always set true.) When a start ID is present on the tape the user bit of the following interface subframe (which is the Q bit of the IU or the first right channel sample of the DAT frame, or R_0) is also set true, and this lasts for 300 ± 30 frames, or about 10 seconds (the same duration as the start ID information on the tape). When a skip ID is encountered in normal play mode (without actually skipping) the user bit of the next left channel subframe (the R bit or L_1) is set true, and this is repeated for 33 ± 3 frames, or about 1 second. When the DAT machine is programmed to act on skip IDs it will skip to the next start ID, and in this case the user bit of the L_1 frame is set true only once – in the first frame that it is encountered. All the other user bits are set to zero.

The number of samples corresponding to a DAT frame depend on the sampling rate, and therefore this dictates the distance between sync, start and skip IDs in the user bits of the interface. At 48 kHz there are 1440 left and right samples per frame, making 2880 subframes between the occurrence of these user bits. At 44.1 kHz this gap is reduced to 2646 words, and in the 32 kHz long play mode found on some players it is 3840 words.

4.8 Channel status data

Channel status data (represented by the C bit in each subframe) is commonly a problematic area with implementations of the standard two-channel interface. It is here that a number of incompatibilities arise between devices and it is here that the main differences exist between professional and consumer formats, because the usage of channel status in consumer and professional equipment is almost entirely different. In this section the principles of channel status usage will be explained, together with an introduction to potential problem areas, although discussion of practical situations is largely reserved until Chapter 7.

4.8.1 Format of channel status

Although there is only one channel status (C) bit in each subframe, these are aggregated over a period of time to form a large data word that contains information about the audio signal being transmitted. The two audio channels theoretically have independent channel status data, although commonly the information is identical, since most applications are for stereo audio. Starting with the frame containing the Z preamble, channel status bits are collected for 192 frames (called a channel status block), resulting in 192 channel status bits for each channel. This long word is subdivided into 24 bytes, each bit of which has a designated function, but the function of these bits depends on whether the application is consumer or professional. The channel status information is updated at block rate, which is 4 ms at a sampling rate of 48 kHz and longer *pro rata* at other sampling rates.

Standards such as AES3 only cover the professional application of channel status, but the EIAJ, IEC and British standards all include a section on consumer applications as well. In the early days of digital interfaces relatively little notice was taken of channel status data by some products and implementations could be sparse, leading to incompatibilities. In AES3 (1992) the format of professional channel status data was extended and explained more carefully, with the intention of setting down more clearly what devices should do with this data if they were to conform properly with the standard. Even more recently the amendments to AES3 have indicated yet further detail in channel status implementation to accommodate such things as new channel modes and higher sampling frequencies. These revisions should ensure that more recent professional devices are more compatible with each other, although it will not help the situation with older equipment. The problem of incompatibility in channel status data is covered further in Chapter 7.

4.8.2   *Professional and consumer usage compared*

It is both a blessing and a curse that the formats of professional and consumer data are so similar. It is a blessing because the user may often wish to transfer material from one to the other and the similarity would appear to make this possible (see section 6.7), but it is a curse because the usage of channel status is so different between the two that the potential for problems is very high, due to the likely misinterpretation of this information by a receiver of the other format. (There is no technical reason, though, why a device should not be designed to interpret both formats correctly.) The first bit of the channel status block indicates whether the usage is consumer (0) or professional (1), and this bit should be interpreted strictly in order to avoid difficulties. (It should be noted that before the consumer format was standardized by the IEC the first bit of the consumer channel status data actually represented 'four-channel mode', not 'consumer/professional'. Some early devices may still exhibit this feature.)

Figure 4.13 shows a graphical comparison of the beginnings of the channel status blocks in professional and consumer modes. Clearly bit 0 is the same and indicates the usage, and bit 1 is also the same, indicating linear PCM audio (0) or 'other purposes' (1) usage of the interface (such as

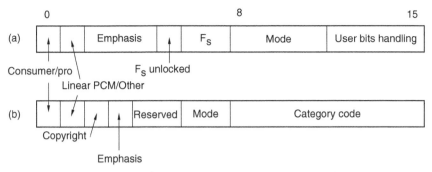

Figure 4.13 Comparison of (a) professional and (b) consumer channel status, bits 1–16.

data-compressed audio, as discussed in section 4.9), but here the similarity stops. In the professional version three bits (bits 2, 3 and 4) are used to signal pre-emphasis, but in the consumer version only two (bits 3 and 4) are used for emphasis indication in the linear PCM mode (although only one is actually used in practice). Bit 2 is used to signify copyright status in the consumer format. Already there is room for incompatibility since a professional device trying to interpret a consumer signal or vice versa could confuse copy protection states with emphasis states. Since copy protection is not normally an issue in professional applications, there is no provision for signalling it in the professional interface.

It is clearly crucial to ensure that the first bit of channel status is correctly interpreted before any further interpretation of channel status takes place. Devices that ignore this fundamental difference in implementation, or that simply ignore most channel status information, may work adequately with other devices some of the time. It is, however, likely that the ever increasing number of possible device configurations will lead to communication difficulties if the channel status implementation is not interpreted strictly.

4.8.3 Professional usage

Usage of channel status and channel modes

Figure 4.14 shows the format of the professional channel status block, and Figures 4.15 and 4.16 show the functions of the first (byte 0) and second (byte 1) bytes in more detail. These indicate basic things about the nature of the source, such as its sampling frequency (a root of problems, as discussed in section 6.7), its pre-emphasis, the mode of operation (mono, stereo, user defined, etc.) and the way in which the user bits are handled

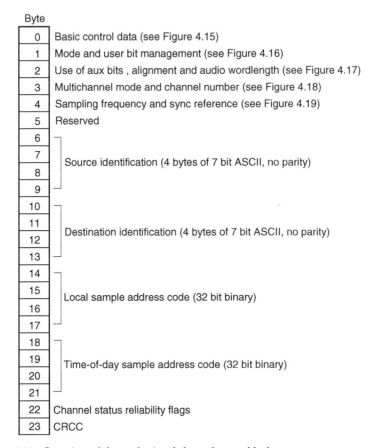

Byte	
0	Basic control data (see Figure 4.15)
1	Mode and user bit management (see Figure 4.16)
2	Use of aux bits , alignment and audio wordlength (see Figure 4.17)
3	Multichannel mode and channel number (see Figure 4.18)
4	Sampling frequency and sync reference (see Figure 4.19)
5	Reserved
6	
7	Source identification (4 bytes of 7 bit ASCII, no parity)
8	
9	
10	
11	Destination identification (4 bytes of 7 bit ASCII, no parity)
12	
13	
14	
15	Local sample address code (32 bit binary)
16	
17	
18	
19	Time-of-day sample address code (32 bit binary)
20	
21	
22	Channel status reliability flags
23	CRCC

Figure 4.14 Overview of the professional channel status block.

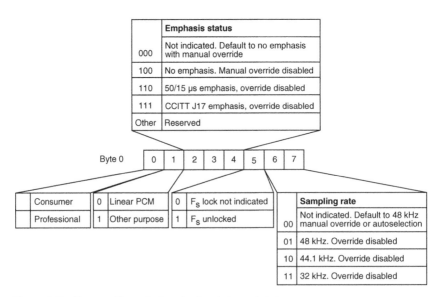

Figure 4.15 Format of byte 0 of professional channel status.

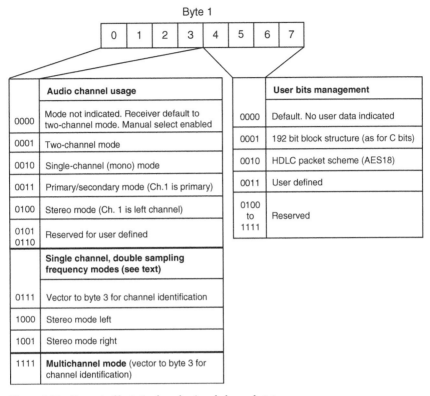

Figure 4.16 Format of byte 1 of professional channel status.

(see section 4.7 and Table 4.1). The sampling frequencies indicated in bits 6–7 of byte 0 are the original and basic ones. If one of the new rates is to be used, as now indicated in byte 4, the byte 0 indication can be set to 00 and the actual rate indicated in byte 4. Indication of the sampling frequency here is not, in any case, a requirement for operation of the interface. Recent versions of the standard use byte 1 to indicate that the interface is operating in single-channel-double-sampling-frequency mode, as discussed in section 4.4. In such a case the sampling frequency indicated in byte 0 is essentially the frame rate of the interface rather than the audio sampling frequency (which would be twice that).

The EBU has made recommendations concerning the primary/ secondary mode of the interface for broadcasting applications in document R72-1999[28]. This identifies three possible uses for the primary and secondary channels as shown in Table 4.3. The means by which these will be indicated elsewhere in channel status has not yet been formally agreed.

Table 4.3 EBU R72 recommendation for use of primary/secondary modes

Primary	Secondary
Complete monophonic mix	Reverse talkback
Mono signal (M)	Stereo difference signal (S)
Commentary	International sound

Use of aux bits and audio resolution

Figure 4.17 shows how byte 2 is split up, with the first three bits describing the use of the auxiliary bits (see section 4.5), and Table 4.4 shows how bits 3–5 are used to indicate audio word length. In Amendment 4 to AES3-1992, bits 6 and 7 are used to indicate the audio alignment level as shown in Table 4.5. Byte 2 allows the source to indicate the number of bits actually used for audio resolution in the main part of the subframe. No matter what the audio resolution, the MSB of the audio sample should always be placed in the MSB position of the interface subframe. This allows receiving devices to adapt their signal processing to handle the incoming signal resolution appropriately, such as when a 16-bit device receives a 20-bit signal, perhaps redithering the audio at the appropriate level for the new resolution in order to avoid distortion. This byte was less comprehensively used in the original version of AES3, only indicating whether the audio word length was maximum 20 bits or 24 bits (in other words, whether the aux bits were available for other purposes or not).

Multichannel modes

Byte 3 is used to describe multichannel modes of the interface, as shown in Figure 4.18. Although the interface can only carry two channels, these may be specific channels of a bundle of associated signals carried over a group of interfaces. The use of the remaining bits depends on the state of bit 7, so that either the whole byte is used to denote the channel number or part of it is used to indicate a particular multichannel mode and the other part to indicate the channel number. The latter is likely to be used for applications like the signalling of particular surround sound configurations, as there are numerous ways in which numbered channels can be assigned to speaker locations (e.g. Left, Right, Centre, Left Surround, Right Surround)[29].

Sampling frequency status

Byte 4 is illustrated in Figure 4.19. The first two bits are used for indicating whether the signal can be used as a sampling frequency reference,

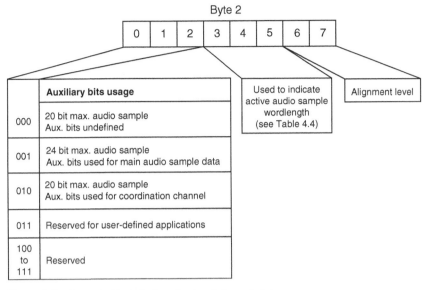

Figure 4.17 Format of byte 2 of professional channel status.

Table 4.4 Use of byte 2 to represent audio resolution

Bits states 3 4 5	Audio word length (24-bit mode)	Audio word length (20-bit mode)
0 0 0	Not indicated	Not indicated
0 0 1	23 bits	19 bits
0 1 0	22 bits	18 bits
0 1 1	21 bits	17 bits
1 0 0	20 bits	16 bits
1 0 1	24 bits	20 bits

Table 4.5 Alignment levels indicated in byte 2, bits 6–7

Bits 6–7	Alignment level
00	Not indicated
01	−20 dB FS (SMPTE RP155)
10	−18.06 dB FS (EBU R68)
11	Reserved

according to the AES11 standard on synchronization. In the '00' state these bits indicate that the signal is not a reference, whilst the '01' state is used to represent a Grade 1 reference and the '10' state is used to represent a Grade 2 reference. This topic is discussed in greater detail in Chapter 6.

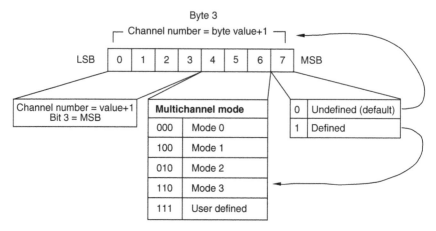

Figure 4.18 Format of byte 3 of professional channel status.

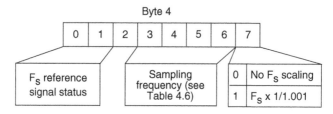

Figure 4.19 Format of byte 4 of professional channel status.

Bits 3–6, in recent amendments, are now used to describe sampling frequencies not originally mentioned in the standard, as detailed in Table 4.6. These are audio sampling frequencies, not necessarily interface frame rates, so they are not dependent on the interface modes described in byte 1. Bit 7 is a sampling frequency scaling flag used to signify (in the '1' state) so-called 'drop-frame' or 'pull-down' frequencies of 1/1.001 times the basic frequency that sometimes arise in post-production operations when digital audio equipment is synchronized to NTSC television signals.

Source and destination identification

In bytes 6 to 13 it is possible to transmit information concerning the source and destination of the signal in the ASCII text format used widely in information technology (see ISO 646). ASCII characters are normally seven bits long (although extended character sets exist which use eight bits), and can represent alphanumeric information. The eighth bit is often used as a parity bit in telecommunications, but it is not used in this

Table 4.6 New sampling frequencies indicated in byte 4, bits 3–6

Bits 6–3 (in that order)	Sampling frequency
0000	Not indicated (default)
0001	24 kHz
0010	96 kHz
0011	196 kHz
1001	22.05 kHz
1010	88.2 kHz
1011	176.4 kHz
1111	User defined
All other states	Reserved

application, being set to zero. (AES3-1985 and IEC 958 specified odd parity, but this was changed in AES3-1992 to no parity.) Some ASCII characters are non-printing symbols called 'control characters' (the first 31 (hex 01–1F) and the last one (hex 7F)) and these are not permitted in this application either. Using this part of channel status the user can 'stamp' the audio signal with a four-character label to indicate the name of the source (bytes 6–9), and the same for the destination (bytes 10–13). It is possible to use destination labelling in automatic routers in order that a signal may control its own routing. The format for these ASCII messages is LSB first, and with the first character of each message in bytes 6 and 10 respectively.

Sample address codes

Bytes 14–21 carry what are called 'sample address codes', which are a form of timecode but counting in audio samples rather than hours, minutes, seconds and frames. Bytes 14–17 are a so-called 'local sample address' which can be used rather like a tape counter to indicate progress through a recording from an arbitrary start point, and bytes 18–21 are a time-of-day code indicating the number of samples elapsed since midnight. Usually this is the time since the device was reset or turned on, because most devices do not have the facility for resetting the sample address code to time of day, but AES3-1992 contains a note to state that it should be the time of day which was laid down during the original source encoding of the signal, and should not be changed thereafter, implying that it should be derived from offtape timecode if that exists.

The four bytes for each sample address code are treated as a 32-bit number, with the LSB sent first; 32 bits allows for a day of 4 294 967 296 samples, which represents just over 24 hours at a sampling rate of 48 kHz. (At the maximum sampling frequency normally allowed for over the interface (54 kHz) 32 bits are not quite enough to represent a whole day's worth of samples, allowing a count of just over 22 hours, but this sampling frequency is normally only used as an upward varispeed of 48 kHz and thus is a non-real-time situation in any case.) If the sampling frequency is known then it is a straightforward matter to convert the sample address to a time of day. The sample address code is updated once per channel status block; thus it is incremented in steps of 4 ms at 48 kHz, and represents the sample address of the first sample of that block.

Byte 22 is used to indicate whether the data contained in certain channel status bytes is reliable, by setting the appropriate bit to '1' in the case of unreliable information, as shown in Table 4.7.

The last byte of the channel status block (byte 23) is a CRC (Cyclic Redundancy Check) designed to detect errors *in channel status information only*, and a simple serial method of producing CRC information is given in AES3-1992. (Most modern AES/EBU interface chips generate channel status CRC automatically.) The CRC byte was not made mandatory in AES3-1985 and a number of systems have not implemented it, but in AES3-1992 it was made mandatory in the 'standard' implementation of the interface (although there is a 'minimum' mode in which it can be left out). The presence or lack of the CRC byte in different devices is a root of incompatibility in older equipment because devices expecting to see it may refuse to interpret channel status data which does not contain a CRC byte and will assume that it is always in error.

Any bits in channel status that are either not used or reserved should be set to the default state of binary '0' – another important factor in ensuring compatibility between devices.

Table 4.7 Use of byte 22 to indicate reliability of channel status data bytes

Bits	Bytes reliable
0–3	Reserved
4	Bytes 0–5
5	Bytes 6–13
6	Bytes 14–17
7	Bytes 18–21

4.8.4 *Levels of professional channel status implementation*

AES3-1985 was rather unclear as to what manufacturers should do with channel status if they were not implementing certain features. Receivers could be found all the way between the two extremes of either interpreting the standard so literally that they expected to see every single bit set 'correctly' before they would work, or else interpreting it so loosely that virtually anything would work, often when it should not. (To be fair it is true that much of the time communication worked without a problem.) In the 1992 revision it was decided to recommend three levels of implementation at the transmitter end, encouraging manufacturers to state the level at which they were working. These levels have been called 'minimum', 'standard' and 'enhanced'. It is intended that at all levels the rest of the frame should be correctly encoded according to the standard.

At the minimum level channel status bits are all set to zero except for the first bit which should be set to signify professional usage. Such an implementation would allow 'belt and braces' communication in most cases, but leaves room for many problems since CRC is not included and neither is any indication of pre-emphasis or sampling frequency. It is intended in such a case that the receiver should set itself to the default conditions (48 kHz, no emphasis, two-channel mode) but allow manual override.

At the standard level the transmitter is expected to implement bytes 0 to 2 and 23 of channel status. This then allows for all the information about the source signal and use of different bits in the frame to be signalled, as well as including the CRC.

An enhanced mode is also allowed which is basically the standard data plus any additional data such as sample address and source identification.

As far as receivers are concerned there is currently little in the way of insistence in the standards concerning how receivers should behave in the case of certain data combinations, except that the manufacturer should state clearly the data recognized and the actions which will be taken. This situation is clarified in AES2-ID, in which a number of equipment classifications are specified, indicating the level to which the device processes and modifies channel status. Manufacturers should publish implementation charts for their devices, much as manufacturers of MIDI-controlled equipment publish such charts in operators' manuals, allowing users to identify the roots of any incompatibility. Such charts, examples of which are found in AES2-ID, should also indicate the treatment of the user bit and validity bit, as well as indicating supported sampling frequencies/resolutions.

In AES2-ID, the implementation of channel status is grouped into A, B and C classes according to the sophistication and nature of the device. Group A devices behave like a wire, simply passing whatever they receive and not storing or acting on it in any way (e.g. a crosspoint switcher). Group B devices all do something with the data they receive, to varying degrees. B1 only decodes audio, not channel status (highly unlikely and possibly risky); B2 decodes channel status but does not store or pass it on; B3 stores and/or passes on channel status data indicated in the implementation chart, modifying it as necessary to reflect the correct state of the audio signal it accompanies. Group C is terminal end equipment (e.g. D/A convertor) that either does not decode channel status (C1) or does and acts on it as indicated (C2).

4.8.5   Overview of channel status in consumer applications

In a way the use of channel status is rather more complicated in consumer applications because of the many types of consumer device and the wide variety of data types that may be transmitted. The format of the basic block is shown in Figure 4.20 and a more detailed breakdown of the first byte (byte 0) is shown in Figure 4.21. (For a comparison with byte 0 of the professional format see Figure 4.15.) Bits 6 and 7 in the consumer format define the 'mode' of use of the channel status block, and so far only mode 0 is standardized with these bits set to '00'. There was a proposal for a mode 1 to be used for 'software information delivery', with the bits set to '10' respectively, in order to allow the transmission of information about production details on prerecorded media. This was written up as a technical report that appears listed as IEC 60958-2 (1994) but is not considered to be formally part of the standard.

The usage of channel status in consumer equipment depends on the mode, and also on the category code which defines the type of device transmitting the data in the second byte (byte 1) of channel status (see the next section). In mode 0, byte 2 of channel status is used to indicate a source number from 1 to 16 and a channel number (see Figure 4.22), so that in the case of sources with multiple audio channels it is possible to signal which two are being transmitted. Byte 3 is used to indicate the sampling frequency of the source and the clock accuracy of the source (see Table 4.8).

Byte 4 of consumer channel status now contains an indication of the audio word length and the original sampling frequency of the signal. It used to be set to all zeros, so older equipment may show this characteristic. Bit 32 indicates whether the sample is maximum 20 or 24 bits

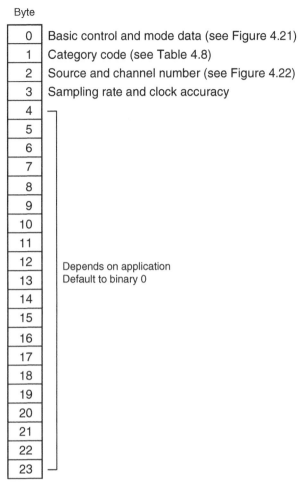

Figure 4.20 Overview of the consumer channel status block.

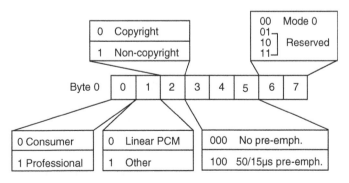

Note: The use of bits 3–5 to signal emphasis and extra (as yet undefined) format information is only relevant in the linear PCM mode of the interface. They are not used for data-reduced 'other' modes in which case they are set to '000'. See IEC 60598 for further details.

Figure 4.21 Format of byte 0 of consumer channel status.

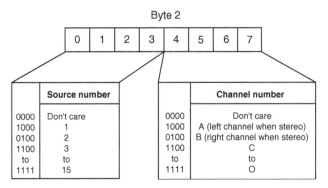

Figure 4.22 Format of byte 2 of consumer channel status.

Table 4.8 Sampling frequency and clock accuracy in
IEC60958 channel status

Bits 24–27	Sampling frequency (kHz)
0000	44.1
0001	88.2
0010	22.05
0011	176.4
0100	48
0101	96
0110	24
0111	192
1000	Not indicated
1100	32
Bits 28–29	*Clock accuracy*
0 0	Level II
0 1	Level III
1 0	Level I
1 1	Frame rate not matched to f_s

(0 and 1 states respectively) and bits 33–35 indicate the resolution in the
same way as the professional interface (Table 4.4). The original sampling
frequency indication in bits 36–39 can be used to show the sampling frequency of a signal before sample rate conversion in a consumer playback
system. This might be the case for applications such as computer games
where sounds with low sampling frequencies could be internally converted
to 44.1 kHz for replay through a common convertor, or with DVD players
where high sampling frequency material is down-converted to 44.1 or
48 kHz for transmission over a standard digital interface. (Currently it is
not permitted to transfer high sampling frequency material over an IEC

60958 digital interface in the DVD standard and digital outputs are limited to basic rates. This is an initial barrier to piracy of high resolution master material, although other procedures are being developed involving watermarking and encryption.)

4.8.6 Category codes in consumer channel status

The eight-bit category code identifies the source device, allowing subsequent channel status and user data to be interpreted correctly. Some examples of the most common category codes are shown in Table 4.9. (Note that category codes are normally written this way round, with the LSB first, which may be confusing since binary numbers are normally written down MSB first.)

As can be seen, all except the General and CD category codes have bit 15 as the 'L' bit. This is used in conjunction with the copyright bit to manage copy protection as explained in the next section. The CD standard was introduced before the copy protection issue became such a 'hot potato' and was stuck with the category code and copyright indication it first used. However, a workaround has been introduced whereby the copyright bit in byte 0 of channel status can be made to alternate states at a rate between 4 and 10 Hz to indicate a 'home copy' of original material of generation 1 or higher.

4.8.7 SCMS and copy protection

The method of coping with copy management is called SCMS (Serial Copy Management System) and is now implemented on all consumer

Table 4.9 Category codes of common products

Category code (bits 8–15; LSB to MSB)	Device type
00000000	General
10000000	CD player
1001001L	MiniDisc
1001100L	DVD
1100000L	DAT
010XXXXL	DSP devices and digital/ digital convertors

digital recording equipment. SCMS applies when signals are copied digitally across a consumer format interface and has no meaning in the professional format. Although SCMS appears not to apply when a signal is copied via an analog interface, in fact there are problems as discussed below.

The principle of SCMS is that a copyright prerecorded signal may be copied only once (provided that the source device is one of the so-called 'white list' of product types from which limited copying is allowed), but no further generations are allowed. This is supposed to allow home users to make a single copy of something they have bought for their own purposes, but to prevent large scale piracy. The copyright protection (Cp) bit that already existed in the consumer format (bit 2) was not sufficient on its own because it did not give any indication of the generation of the copy. The so-called 'L bit' was therefore introduced in the category code (see previous section). The Cp bit in conjunction with the L bit can be used to prevent more than one generation of copying of copyright material. The state of the L bit in effect signifies the 'generation' of the signal, whether '0th generation' (an original prerecorded work) or 1st generation and higher copies. The Cp bit now signifies whether the source material is copyright or not ('0' = copyright). Unfortunately the complication does not stop here, because although the L bit is normally set so that in the '1' state it represents an original (commercially released, prerecorded software) rather than a copy, with laser optical products and broadcast receivers it is the other way around!

The upshot of all this is that if a recording device sees that the C bit of a digital source is '0' (copyright material) and the L bit indicates that the source is original prerecorded material, it will allow the copy. If the L bit indicates that the source is already a copy it will disallow the recording. When copyright material is copied from a prerecorded source a flag is recorded on the copy to state 'I am a copy of a copyright source', and when this recording is replayed the L bit will be set to show that it is not an original, thus disallowing further copies. Extremely thorough coverage of SCMS is to be found in an AES journal article by Sanchez[30].

4.8.8   SCMS in DAT machines

SCMS was first introduced because of the perceived threat of copying with DAT machines. It works in such machines as follows. There are two DAT category codes: one called simply 'DAT' (code 11000000) and one called 'DAT-P' (code 11000001) – the difference between them is the L bit.

On DAT recordings there are also two bits in the subcode recorded on the tape that controls copy protection and these are called the 'ID6' bits. An SCMS DAT machine looks at the ID6 bits recorded on the tape it is playing to determine how it should set the combination of L and C bits on the digital interface.

If the ID6 on the source tape is 00 (copies allowed) it will set the category to 'DAT' and the C bit to show 'non-copyright'. An SCMS machine receiving that signal would also set the recorded ID6 of the copy to 00, since there would be no reason to prevent further copies, and any number of serial digital copies might then be made.

If a DAT recorder sees a 'DAT-P' source it will allow copies no matter what the status of the C bit, whereas if it sees a straightforward 'DAT' source it will only allow a copy if the source is not copy protected. So if the ID6 on the source tape is 10 (copies not allowed), the source machine will set the category to 'DAT' and the C bit to 'copyright', then the recorder will not be able to copy that tape. If the ID6 on the source tape is 11 (one copy allowed), the machine will set the category to 'DAT-P' and the C bit to 'copyright', then a receiver would be able to record the signal (since DAT-P allows copies no matter what the © status). It would know, though, that it was copying © material. The ID6 of the copy would then automatically be set to 10 to prevent further copies.

Concerning copies made between DAT machines and other systems, or between pre-SCMS and SCMS machines, the following applies:

1 Digital copies from CD

The CD category code (10000000) when recognized by an SCMS DAT machine will result in a copy whose ID6 is set to 10 (copies not allowed). (The L bit is '0' for original prerecorded material in laser optical products.)

2 Digital copies from other digital sources

Copies made from sources having the 'General' category code (00000000) will have their ID6 set to 11, whatever the © status, allowing one further copy only. Sources asserting this code are likely to be such things as A/D convertors and some older DAT machines. It acknowledges that the source of the material is unclear, and might be copyright or might not.

3 Digital copies to SCMS DAT machines from pre-SCMS machines

Pre-SCMS machines may use either the 'General' category or the 'DAT' category, depending on when they were made and by whom. They will *not* normally be able to recognize the difference between a recorded

ID6 of 11 and an ID6 of 10, because prior to SCMS the machine only had to look at one bit to detect © status. Therefore such a machine will normally interpret both codes as indicating that the recording is copy protected (not even allowing one copy), and set the © flag on the digital output.

Whether or not the receiver will record the data depends on whether the category is 'General' or 'DAT'. If it is 'General' then see 2 above. If it is 'DAT', then not even a single copy will be allowed. The only case in which unlimited copies will be allowed is when the source tape has an ID6 of 00, which is likely to be the case with many tapes recorded on pre-SCMS machines.

4 Digital copies of recordings made from analog inputs

Unfortunately, SCMS DAT machines will set the ID6 of analog-sourced recordings to 11, thus allowing only one digital copy. This is a nuisance when the source is a perfectly legitimate non-copyright signal, such as one of your own private recordings.

5 Digital copies of prerecorded DAT tapes

The ID6 of prerecorded tapes is set to 11, thus allowing one further copy if using an SCMS machine. If using a pre-SCMS replay machine, the 11 will be interpreted as 10 (see 3 above) and the © bit will be asserted on the interface. A copy will only be possible if the category of the source machine is 'General', but not if it is 'DAT'.

6 Recording non-copy-protected material on SCMS machines

There is no way to record completely unprotected material on an SCMS machine, except by feeding it with a digital source having a category code other than 'General' and a recorded ID6 of 00. This might be feasible if you have an early DAT machine. Even material recorded via the SCMS machine's analog inputs cannot be copied beyond a single generation.

7 Digital copying from SCMS machines to pre-SCMS machines

Such copies will only be possible at 48 kHz (or 32 kHz if you have such a tape); 44.1 kHz recordings will be blocked on unmodified machines. Source tapes with ID6 set to either 11 or 10 will cause the CP status to be asserted on the digital interface, and, since pre-SCMS machines tend to ignore the L bit, the copy will not be allowed at all. Source tapes with ID6 set to 00 may be copied.

8 Manual setting of ID6 status

Consumer machines will not allow the ID6 status of tapes to be set, but some recent professional machines will allow this.

4.8.9 Channel status in consumer CD machines

It was originally intended that the first four bits of the Q subcode from the CD would be copied into the first four bits of the channel status data. The first four bits of Q subcode are as follows:

Bit 0   Two or four channel (two channel = '0')
Bit 1   Undefined
Bit 2   Copy protect
Bit 3   Pre-emphasis

Apart from bit 0 these are compatible with IEC 60958. Since CDs have never implemented a four-channel mode, bit 0 remains in the '0' state which is compatible with the 'consumer' status of bit 0 of IEC 60958. Other than this, the channel status format of the CD category code is the same as the general format, with the sampling frequency bits set to '0000' to indicate 44.1 kHz. The user bits contain the subcode information as discussed in section 4.7.2.

4.9 Data-reduced audio over standard two-channel interfaces

4.9.1 General principles

The standard two-channel interface was originally designed for linear PCM audio samples but in recent years there has been increasing use of data-reduced audio coding systems such as Dolby Digital (AC-3), DTS and MPEG. Because consumer systems in particular needed the ability to transfer such signals digitally, the non-audio mode of the interface has been adapted to the purpose. This is described in a relatively new IEC standard numbered 61937[31]. In addition to a general specification detailing the principles it also has a number of parts that describe the handling of specific data-reduced audio formats, some of which are not yet finalized at the time of writing. A similar but not identical SMPTE standard (337M)[32] describes professional non-audio applications of the interface, including its use for carrying Dolby E data (see below). SMPTE 338M and 339M specify data types to be used with this standard. The SMPTE standard is also more generic than the IEC standard, designed to deal with a variety of data uses of the interface, not just low-bit rate audio. It can also carry time-stamp data in the form of SMPTE 12M timecode. The reader is referred to the standards for more precise details regarding implementation of specific formats.

In both SMPTE and IEC versions the low-bit rate audio data is carried in bursts in place of the normal linear PCM audio information, with bit 1 of channel status set to the 'other uses' or 'non-audio' state. (The SMPTE standard makes it clear, though, that professional devices should not rely on this. Some digital video tape recorders, for example, cannot control the non-audio bit yet it is desirable that they should receive/transmit and store low-bit rate audio such as Dolby E.) Sometimes the validity bit is also set to indicate 'invalid' or 'unsuitable for conversion to analog', as a further measure. In the IEC standard the data is carried in the 16 most significant bits of the audio data slot, from bits 12 to 27 in IEC nomenclature. In the SMPTE standard it is possible for the data to occupy 16, 20 or 24 bits. In the IEC standard the two interface subframes are treated as conveying a single data stream whereas in the SMPTE standard the subframes can be handled together or separately (for example, one subframe could carry PCM and the other data-reduced audio).

A data burst of low-bit rate audio (representing the encoded frame of a number of original PCM samples) typically occupies a number of consecutive subframes, the last subframe of the burst being packed with zeros if required. Each burst is preceded by a preamble of four 16-bit words, the first two of which are a synchronization pattern, the third of which indicates the mode of data being carried and the fourth of which indicates the length of the burst, as shown in Figure 4.23. In the IEC standard up to eight independent bitstreams can be carried in this way, each being identified in the third byte of the preamble. The SMPTE standard can carry up to 14 independent streams in the independent subframe mode and the data type preambles are not identical to IEC 61937. Because the total data rate may be lower than that required for linear PCM audio, packing zero bits can be used between bursts of low-bit rate audio data and there is a requirement for at least four subframes to have bits 12–27 set to zero every 4096 frames.

4.9.2 Data-reduced consumer formats

A number of parts of IEC 61937 describe the transmission of audio signals encoded to different standards, some of which are manufacturer- or

Figure 4.23 Format of the data burst in IEC 61937.

system-specific and others are internationally standardized. Data-reduced bitstreams of current relevance here are AC-3 (Dolby Digital), DTS (Digital Theatre Systems), MPEG 1 and 2-BC, MPEG 2-AAC and Sony ATRAC. The most commonly encountered applications in consumer systems at the moment are for the transfer of encoded multichannel surround sound data from DVD players to home cinema systems, using either Dolby Digital or DTS encoding. The digital output of DVD players is typically an IEC 60958 interface on a phono connector or optical interface that can be used to carry 5.1-channel surround sound data for decoding and D/A conversion in a separate surround sound processor and amplifier.

4.9.3 Data-reduced professional formats

The SMPTE 337M standard allows a number of data-reduced audio formats to be transmitted over the interface. The most commonly used of these are Dolby AC-3 and Dolby E, but data types are also specified for MPEG 1 and 2.

Dolby E is a data reduction system designed for professional purposes, using mild data reduction in order to minimize generation losses. It was introduced to satisfy a need to transfer production multichannel surround sound signals over two-channel media such as digital interfaces and video tape recorder audio tracks, in order to ease the transition from two-channel operations to 5.1-channel operations in broadcasting and post-production environments. It packs the audio data into the two-channel frame in a similar way to that described in the previous section. The resolution can be adapted to fit 16-, 20- or 24-bit media, the most common implementation using 20-bit frame format mode (both subframes used together) at a data rate, including overheads, of about 1.92 Mbit/s. (The 16- and 24-bit modes run at data rates of 1.536 and 2.304 Mbit/s respectively.) Dolby E packets are aligned with video frames so that the audio can be switched or edited synchronously with video. For example, there are 25 Dolby E packets per second when synchronized with 25 fps video.

4.10   AES42 digital microphone interface

This digital microphone interface is based on the AES3 two-channel interface and includes options for powering and synchronization of microphones. Most digital microphones currently in existence employ conventional capsule technology with A/D conversion very close to the microphone capsule rather than direct digital transduction of the acoustic waveform. There are nonetheless some advantages to be had in using digital transmission of audio signals from microphones, principally the potential for higher quality and lower noise as a result of conversion close to the capsule and the avoidance of long cable runs in the analog domain at low signal level. A microphone that conforms to this standard is typically referred to as an AES3-MIC.

The AES42 standard[33] notes that some patent rights may relate to the interface in question and that licensing of some elements may be required.

4.10.1 Principles

The AES42 (AES3-MIC) interface is a standard AES3 interface that also carries power for the microphone. There is also a proposal to adopt a slightly different XLR connector to the normal one, termed the XLD connector, intended to avoid the possibility of damaging equipment not designed for the power supplying capacity of this interface. The XLD connector is identified with a striped 'zebra' ring to distinguish it, but this is not mandatory and there is some disagreement about the need for it (some say that studio practice has managed adequately for years with some XLR connectors carrying phantom power and others not). A combination of coded grooves and keys enables XLD connectors to be used in a variety of combinations with ordinary XLR connectors, or in the fully coded form may prevent one connector from being inserted into a socket of the other type.

If the microphone is monophonic both subframes of the digital interface carry identical information, except in single-channel-double-sampling-frequency modes where they carry successive samples of the one channel.

4.10.2 Powering

The form of phantom powering in this standard is not the same as the 48 volt system used with analog microphones. In this standard the so-called 'digital phantom power' (DPP) is 10 volts applied to both legs of the balanced AES3 cable via a centre tap on the cable side of the transformer. Maximum continuous load is specified as 250 mA, with a peak load of 300 mA when charging additional load capacitance.

4.10.3 Remote control and status reporting

An AES3-MIC may be remote controlled using pulsed modulation of the power supply voltage (see below), with positive-going pulses of 2 ± 0.2 volts that carry data at a rate of 750 bits per second (at 48 kHz sampling frequency or multiples) or proportionally lower for 44.1 kHz and multiples. The remote control information can indicate changes of microphone settings such as directivity pattern (omni, cardioid, etc.), attenuation, limiting, gain, muting and high-pass filtering. There is also the option for manufacturer-specific settings and extended instructions involving changes of more sophisticated features such as dither type, sampling frequency and so forth.

The microphone's status can be reported back to the receiver by means of the user bit channel of the AES3 interface. In this application the user bits are assembled into a 24-byte structure, in the same way as channel status information, synchronized by the same Z preamble that indicates the start of a 192-bit channel status block.

4.10.4 Synchronization

There are two modes of operation of an AES3-MIC. In Mode 1 the microphone is self-clocking and generates its own sampling frequency reference. As a consequence of this all mics in a studio would be unsynchronized and each would have a slightly different sampling frequency. Any mixing console dealing with their digital signals would have to apply sampling frequency conversion. In Mode 2, microphones can be synchronized to a common reference signal and this is achieved by transmitting additional data in the remote control information to the microphone. As shown in Figure 4.24 it is intended that a phase comparator be present in each AES3-MIC receiver (say a mixing console input) that compares the relative phase of the word clock extracted from the incoming microphone

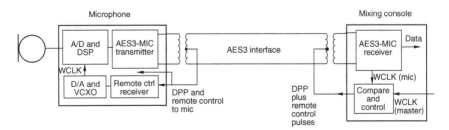

Figure 4.24 Conceptual example of AES42 digital microphone interface.

signal to a reference signal. A binary value is returned to the microphone in the remote control data that adjusts the frequency of its internal clock accordingly, using a D/A convertor to convert the remote control data into a DC voltage and a voltage-controlled crystal oscillator to generate the word clock.

The resolution of the sync information can be extended from eight to 13 bits for use in high resolution applications where clock accuracy and low jitter are crucial.

4.11 The standard multichannel interface (MADI)

Originally proposed in the UK in 1988 by four manufacturers of professional audio equipment (Sony, Neve, Mitsubishi and Solid State Logic), the so-called 'MADI' interface is now an AES and ANSI standard. It was designed to simplify cabling in large installations, especially between multitrack recorders and mixers, and has a lot in common with the format of the two-channel interface. The standard concerned is AES10-1991[12] (ANSI S4.43-1991), and a recent draft revision has been issued. This interface was intentionally designed to be transparent to standard two-channel data making the incorporation of two-channel signals into a MADI multiplex a relatively straightforward matter. The original channel status, user and auxiliary data remain intact within the multichannel format.

MADI stands for Multichannel Audio Digital Interface; in the original standard 56 channels of audio are transferred serially in asynchronous form and consequently the data rate is much higher than that of the two-channel interface. For this reason the data is transmitted either over a coaxial transmission line with 75 ohm termination (not more than 50 m) or over a fibre optic link. The protocol is based closely on the FDDI (Fibre-Distributed Digital Interface) protocol suggesting that fibre optics would be a natural next step[34]. The recent draft revision proposes a means of allowing higher sampling frequencies and an extension of the channel capacity.

4.11.1 Format of the multichannel interface

The serial data structure is as shown in Figure 4.25. It is divided into subframes which, apart from the preamble area, are identical to AES3 subframes. The preamble is not required here because the interface is synchronized in a different way, so the four-bit slot is replaced with four 'mode bits' the functions of which are labelled in the diagram. There are

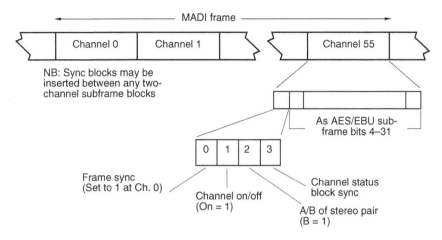

Figure 4.25 Format of the MADI frame.

56 subframes in a frame and bit 0 signifies the start of channel 0 (it is set true for that frame only); bit 1 indicates whether a particular subframe or audio channel is active (1 for active); bit 2 indicates whether the subframe is either the A or B channel of a two-channel pair derived from an AES3 source (1 for 'B'); and bit 3 indicates the start of a new channel status block for the channel concerned. The audio part of the frame is handled in the same way as in the two-channel interface, and the V, U, C and P bits apply on a per-channel basis, with parity applying over bits 4–31.

The channel code is different from that used in the two-channel version, and another important contrast is that the link transmission rate is independent of the audio sampling frequency or number of channels involved. In the original standard the highest data transfer rate is at the highest sampling rate (54 kHz) and number of channels (56) times the number of bits per subframe (32), that is $54\,000 \times 56 \times 32 = 96.768\,\text{Mbit/s}$. It is assumed that the transmitter and receiver will be independently synchronized to a common sampling frequency reference in order that they operate at identical sampling frequencies. In the recent draft revision sampling frequencies up to 96 kHz are allowed and the maximum channel capacity is extended to 64 by limiting the sampling frequency to no more than 48 kHz (removing the varispeed tolerance in other words). Samples of 96 kHz are handled by reducing the channel capacity to 28 and by either using an approach similar to the AES3 single-channel-double-sampling-frequency mode described earlier, or by transmitting two sets of samples successively within one 20.8 µs frame.

The MADI link itself does not synchronize the receiver's sampling clock. The channel-coding process involves two stages: first the 32-bit subframe is divided into groups of 4 bits, and these groups are then encoded into 5-bit words chosen to minimize the DC content of the data signal, according to Table 4.10 (4/5-bit encoding).

The actual *transmission rate* of the data is thus 25% higher than the original data rate, and 32-bit subframes are transmitted as 40 bits. To carry the 4/5-encoded data over the link a '1' is represented by a transition (in either direction) and a '0' by no transition, as shown in the example of Figure 4.26.

Special synchronization symbols are inserted by the transmitter in between encoded subframes, and these take the binary form 11000 10001, transmitted from the left (a pattern which does not arise otherwise). These have the function of synchronizing the receiver (but not its sample clock) and may be inserted between subframes or at the end of the frame in

Table 4.10   4/5-bit encoding in MADI

4-bit groups	5-bit codes
0000	11110
0001	01001
0010	10100
0011	10101
0100	01010
0101	01011
0110	01110
0111	01111
1000	10010
1001	10011
1010	10110
1011	10111
1100	11010
1101	11011
1110	11100
1111	11101

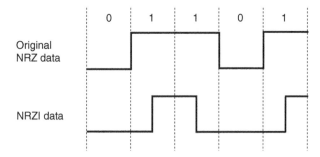

Figure 4.26   An example of the NRZI channel code.

order to fill the total data capacity of the link which is 125 Mb/s ± 100 ppm. The prototype MADI interfaces were designed around AMD's TAXI (Transparent Asynchronous Xmitter/Receiver interface) chips, which were becoming more widely used in high speed computer networks, and these chips normally take care of the insertion of synchronizing symbols so that the transmission rate of the link remains constant.

4.11.2 Electrical characteristics

The coaxial version of the interface consists of a 75 ohm transmission line terminated in BNC connectors, using cable with a characteristic imped-ance of 75 ± 2 ohms and losses of <0.1 dB/m (1–100 MHz). Suggested driver and receiver circuits are illustrated in the standard, and the receiver is expected to decode data with a minimum eye pattern as shown in Figure 4.27. Equalization is not permitted at the receiver, and distances of up to 50 metres may be covered.

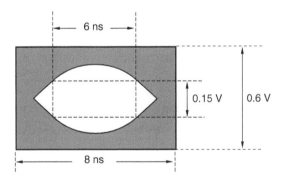

Figure 4.27 The minimum eye pattern acceptable for correct decoding of MADI data.

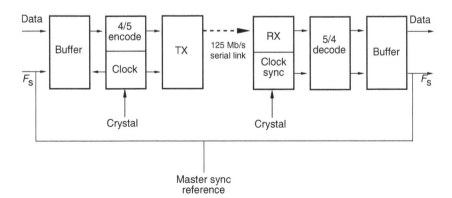

Figure 4.28 Block diagram of MADI transmission and reception.

The block diagram of transmitter-to-receiver communication is shown in Figure 4.28. Here the source data is buffered, 4/5 encoded and then formatted with the sync symbols before being transmitted. The receiver extracts the sync symbols, decodes the 4/5 symbols and a buffer handles any short-term variation in timing due to the asynchronous nature of the interface. An external sync reference ensures that the two sampling clocks are locked.

4.12   Manufacturer-specific interfaces

Interfaces other than the internationally standardized two-channel and multichannel types will be described and discussed in this section. For example, a number of interfaces have been introduced by specific manufacturers and are normally only found on that manufacturer's products, or are licensed for use by others. Some of these proprietary interfaces have become quite widely used in commercial products – for example, the ADAT 'lightpipe' interface is widely encountered on computer sound cards because it is a small-sized optical connector capable of carrying eight channels of digital audio. Some of the technology described in this chapter is the subject of patents and implementers may need to enter into licensing agreements.

MIDI, included briefly in the previous editions of this book, is not strictly a digital audio interface and coverage of it has been removed in this edition. The interested reader is referred to *MIDI Systems and Control*[35] or the forthcoming *Desktop Audio Technology* by Francis Rumsey.

4.12.1   Sony digital interface for LPCM (SDIF-2)

Sony's original interface for linear PCM data was SDIF-2. It was designed for the transfer of one channel of digital audio information per physical link at a resolution of up to 20 bits (although most devices only make use of 16). The interface has also been used on equipment other than Sony's, for the sake of compatibility, but the use of this interface is declining as the standard two-channel interface becomes more widely used.

The interface is unbalanced and uses 75 ohm coaxial cable terminating in 75 ohm BNC-type connectors, one for each audio channel. TTL-compatible electrical levels (0–5 V) are used. The audio data is accompanied by a word clock signal on a separate physical link (see Figure 4.29), which is a square wave at the sampling frequency used to synchronize the receiver's sample clock. Sony's multitrack machines use SDIF also, but

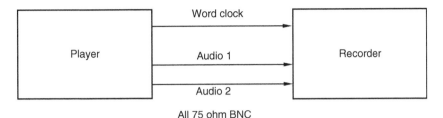

Figure 4.29   SDIF-2 interconnection for two audio channels.

Table 4.11   Pinouts for differential SDIF
multichannel interface

Pin	*Function*
1, 2	Ch. 1 $(-/+)$
3, 4	Ch. 2 $(-/+)$
5, 6	Ch. 3 $(-/+)$
etc.	etc.
to	to
47, 48	Ch. 24 $(-/+)$
49, 50	NC

with a differential electrical interface conforming to RS-422 standards
(see section 1.7.2) and using 50 pin D-type multiway connectors, the
pinouts of which are shown in Table 4.11. A single BNC connector carries
the word clock as before.

In each audio sample period, the equivalent of 32 bits of data is transmitted over each physical link, although only the first 29 bits of the word
are considered valid, since the last three-bit cell periods are divided into
two cells of one-and-a-half times the normal duration, violating the NRZ
code in order to act as a synchronizing pattern. As shown in Figure 4.30,
20 bits of audio data are transmitted with the MSB first (although
typically only 16 bits are used), followed by nine control or user bits
(although the user bits are rarely employed). A block structure is created
for the control/user bits which repeats once every 256 sample periods,
signalled using the block sync flag in bit 29 of the first word of the block.
The resulting data rate is 1.53 Mb/s at 48 kHz sampling rate and
1.21 Mb/s at 44.1 kHz.

The SDIF-2 interface was originally used mainly for the transfer of
audio data between Sony professional digital audio products, particularly
the PCM-1610 and 1630 PCM adaptors, but also from semi-professional
Sony equipment which had been modified to give digital inputs and

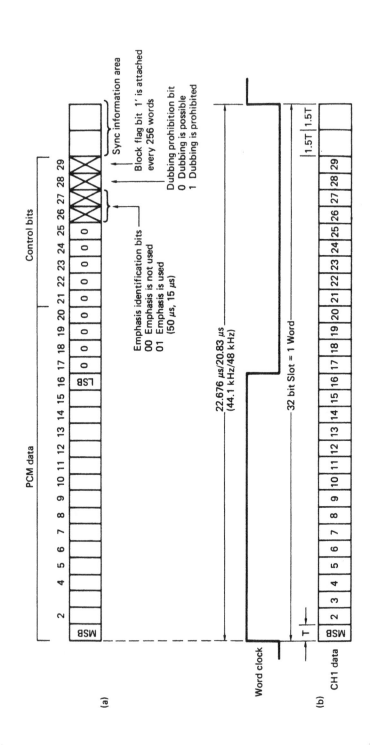

(a)

PCM data

Control bits

MSB

LSB

2 4 5 6 7 8 9 10 11 12 13 14 15 16 17 18 19 20 21 22 23 24 25 26 27 28 29

Emphasis identification bits
00 Emphasis is not used
01 Emphasis is used
(50 μs, 15 μs)

Dubbing prohibition bit
0 Dubbing is possible
1 Dubbing is prohibited

Block flag bit '1' is attached
every 256 words

Sync information area

Word clock

22.676 μs/20.83 μs
(44.1 kHz/48 kHz)

32 bit Slot = 1 Word

T

(b)

CH1 data

MSB

2 3 4 5 6 7 8 9 10 11 12 13 14 15 16 17 18 19 20 21 22 23 24 25 26 27 28 29

1.5T 1.5T

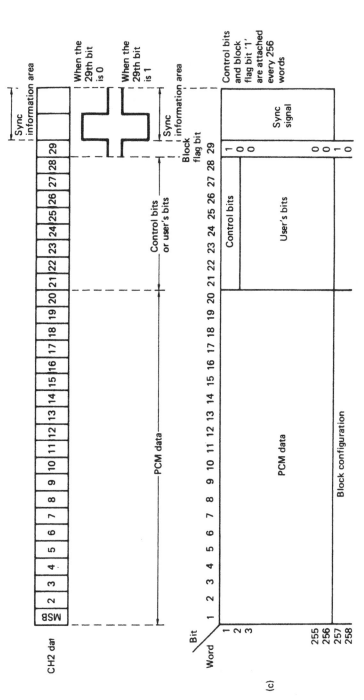

Figure 4.30 At (a) is the clock content of the SDIF signal; (b) shows the synchronizing pattern used for data reception. At (c) user bits form a block that is synchronized every 256 sample periods.

outputs (such as the PCM-701 and various DAT machines). It has also been used on a number of disk-based workstations. It is not recommended for use over long distances and it is important that the coaxial leads for each channel and the word clock are kept to the same length otherwise timing errors may arise. Problems occasionally arise with third-party implementations of this interface that do not use the 1.5-bit cell sync pattern at the end of words, requiring some trial and error involving delays of the data signal with relation to the separate word clock in order for the link to function correctly.

4.12.2 Sony digital interface for DSD (SDIF-3)

Sony has recently introduced a high-resolution digital audio format known as 'Direct Stream Digital' or DSD. This encodes audio using one-bit sigma–delta conversion at a very high sampling frequency of typically 2.8224 MHz (64 times 44.1 kHz). There are no internationally agreed interfaces for this format of data, but Sony has released some preliminary details of an interface that can be used for the purpose, known as SDIF-3. Some early DSD equipment used a data format known as 'DSD-raw' which was simply a stream of DSD samples in non-return-to-zero (NRZ) form, as shown in Figure 4.31(a).

In SDIF-3 data is carried over 75 ohm unbalanced coaxial cables, terminating in BNC connectors. The bit rate is twice the DSD sampling frequency (or 5.6448 Mbit/s at the sampling frequency given above) because phase modulation is used for data transmission as shown in Figure 4.31(b). A separate word clock at 44.1 kHz is used for synchronization purposes. It is also possible to encounter a DSD clock signal connection at 64 times 44.1 kHz (2.8224 MHz).

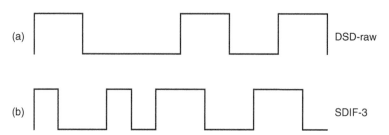

Figure 4.31 Direct Stream Digital interface data is either transmitted 'raw' as shown at (a) or phase modulated as in the SDIF-3 format shown at (b).

4.12.3   Sony multichannel DSD interface (MAC-DSD)

Sony has also developed a multichannel interface for DSD signals, capable of carrying 24 channels over a single physical link[36]. The transmission method is based on the same technology as used for the Ethernet 100BASE-TX (100 Mbit/s) twisted-pair physical layer (PHY), but it is used in this application to create a point-to-point audio interface. Category 5 cabling is used, as for Ethernet, consisting of eight conductors. Two pairs are used for bi-directional audio data and the other two pairs for clock signals, one in each direction.

Twenty-four channels of DSD audio require a total bit rate of 67.7 Mbit/s, leaving an appreciable spare capacity for additional data. In the MAC-DSD interface this is used for error correction (parity) data, frame header and auxiliary information. Data is formed into frames that can contain Ethernet MAC headers and optional network addresses for compatibility with network systems. Audio data within the frame is formed into 352 32-bit blocks, 24 bits of each being individual channel samples, six of which are parity bits and two of which are auxiliary bits.

In a recent enhancement of this interface, Sony has introduced 'SuperMAC' which is capable of handling either DSD or PCM audio with very low latency (delay), typically less than 50 μs. The number of channels carried depends on the sampling frequency. Twenty-four DSD channels can be handled, or 48 PCM channels at 44.1/48 kHz, reducing proportionately as the sampling frequency increases. In conventional PCM mode the interface is transparent to AES3 data including user and channel status information.

4.12.4   Tascam digital interface (TDIF)

Tascam's interfaces have become popular owing to the widespread use of the company's DA-88 multitrack recorder and more recent derivatives. The primary TDIF-1 interface uses a 25 pin D-sub connector to carry eight channels of audio information in two directions (in and out of the device), sampling frequency and pre-emphasis information (on separate wires, two for f_s and one for emphasis) and a synchronizing signal. The interface is unbalanced and uses CMOS voltage levels. Each data connection carries two channels of audio data, odd channel and MSB first, as shown in Figure 4.32. As can be seen, the audio data can be up to 24 bits long, followed by two bits to signal the word length, one bit to signal emphasis and one for parity. There are also four user bits per channel that are not usually used. This resembles a modified form of the AES3 interface frame

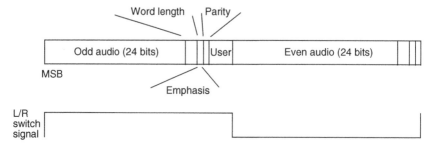

Figure 4.32 Format of TDIF data and LRsync signal.

format. An accompanying left/right clock signal is high for the odd samples and low for the even samples of the audio data. It is difficult to find information about this interface but the output channel pairs appear to be on pins 1–4 with the left/right clock on pin 5, while the inputs are on pins 13–10 with the left/right clock on pin 9. Pins 7, 14–17 (these seem to be related to output signals) and 22–25 (related to the input signals) are grounded. The unbalanced, multi-conductor, non-coaxial nature of this interface makes it only suitable for covering short distances up to 5 metres.

4.12.5 *Alesis digital interface*

The ADAT multichannel optical digital interface, commonly referred to as the 'light pipe' interface or simply 'ADAT Optical', is a serial, self-clocking, optical interface that carries eight channels of audio information. It is described in US Patent 5,297,181: 'Method and apparatus for providing a digital audio interface protocol'. The interface is capable of carrying up to 24 bits of digital audio data for each channel and the eight channels of data are combined into one serial frame that is transmitted at the sampling frequency. The data is encoded in NRZI format for transmission, with forced ones inserted every five bits (except during the sync pattern) to provide clock content. This can be used to synchronize the sampling clock of a receiving device if required, although some devices require the use of a separate 9 pin ADAT sync cable for synchronization. The sampling frequency is normally limited to 48 kHz with varispeed up to 50.4 kHz and TOSLINK optical connectors are typically employed (Toshiba TOCP172 or equivalent). In order to operate at 96 kHz sampling frequency some implementations use a 'double-speed' mode in which two channels are used to transmit one channel's audio data (naturally halving the number of channels handled by one serial interface).

Figure 4.33 Basic format of ADAT data.

Although 5 metre lengths of optical fibre are the maximum recommended, longer distances may be covered if all the components of the interface are of good quality and clean. Experimentation is required.

As shown in Figure 4.33 the frame consists of an 11-bit sync pattern consisting of 10 zeros followed by a forced one. This is followed by four user bits (not normally used and set to zero), the first forced one, then the first audio channel sample (with forced ones every five bits), the second audio channel sample, and so on.

4.12.6 Roland R-bus

Roland has recently introduced its own proprietary multichannel audio interface that, like TDIF (but not directly compatible with it), uses a 25-way D-type connector to carry eight channels of audio in two directions. Called R-bus it is increasingly used on Roland's digital audio products, and convertor boxes are available to mediate between R-bus and other interface formats. Little technical information about R-bus is available publicly at the time of writing.

4.12.7 Mitsubishi digital interfaces

This section is included primarily for historical completeness, as Mitsubishi no longer manufactures digital audio equipment. Mitsubishi's ProDigi format tape machines used a digital interface similar to SDIF but not compatible with it. Separate electrical interconnections were used for each audio channel. Interfaces labelled 'Dub A' and 'Dub B' were 16-channel interfaces found on multitrack machines, handling respectively tracks 1–16 and 17–32. These interfaces terminated in 50-way D-type connectors and utilized differential balanced drivers and receivers. One sample period was divided into 32-bit cells, only the first 16 of which were used for sample data (MSB first), the rest being set to zero. There was no sync pattern within the audio data (such as there is between bits 30 and 32 in the SDIF-2 format). The audio data was accompanied by a separate bit clock (1.536 MHz square wave at 48 kHz sampling rate) and a word clock

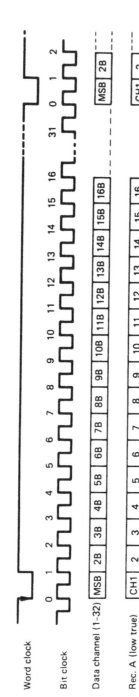

Figure 4.34 Data format of the Mitsubishi multitrack interface.

signal which went low only for the first bit cell of each 32-bit audio data word (unlike SDIF which uses a sampling rate square wave), as shown in Figure 4.34. Status information was passed over two separate channels, which take the same format as an audio channel but carried information about the record status of each of the 32 channels of the ProDigi tape machine. One status channel (Rec 'A') handled tracks 1–16, and the other (Rec 'B') handled tracks 17–32 of a multitrack machine. The pin assignments for these connectors are shown in Table 4.12.

Mitsubishi interfaces labelled 'Dub C' were two-channel interfaces. These terminated in 25-way D-type connectors and utilized unbalanced drivers and receivers. One sample period was divided into 24-bit cells, only the first 16 or 20 of which are normally used, depending on the resolution of the recording in question. Again audio data is accompanied by

Table 4.12(a)   Pinouts for Mitsubishi 'Dub A' connector

Pin	Function	Pin	Function
1, 18	Ch. 1 (+/−)	10, 27	Ch. 10 (+/−)
2, 19	Ch. 2 (+/−)	11, 28	Ch. 11 (+/−)
3, 20	Ch. 3 (+/−)	12, 29	Ch. 12 (+/−)
4, 21	Ch. 4 (+/−)	13, 30	Ch. 13 (+/−)
5, 22	Ch. 5 (+/−)	14, 31	Ch. 14 (+/−)
6, 23	Ch. 6 (+/−)	15, 32	Ch. 15 (+/−)
7, 24	Ch. 7 (+/−)	16, 33	Ch. 16 (+/−)
8, 25	Ch. 8 (+/−)	17, 50	GND
9, 26	Ch. 9 (+/−)	34, 35	Bit clock (+/−)
36, 37	WCLK (+/−)	38, 39	Rec A (+/−)
40, 41	Rec B (+/−)		

Table 4.12(b)   Pinouts for Mitsubishi 'Dub B' connector

Pin	Function	Pin	Function
1, 18	Ch. 17 (+/−)	9, 26	Ch. 25 (+/−)
2, 19	Ch. 18 (+/−)	10, 27	Ch. 26 (+/−)
3, 20	Ch. 19 (+/−)	11, 28	Ch. 27 (+/−)
4, 21	Ch. 20 (+/−)	12, 29	Ch. 28 (+/−)
5, 22	Ch. 21 (+/−)	13, 30	Ch. 29 (+/−)
6, 23	Ch. 22 (+/−)	14, 31	Ch. 30 (+/−)
7, 24	Ch. 23 (+/−)	15, 32	Ch. 31 (+/−)
8, 25	Ch. 24 (+/−)	16, 33	Ch. 32 (+/−)
17, 50	GND		

Table 4.13 Pinouts for Mitsubishi 'Dub C' connectors

Pin	Function
1, 14	Left $(+/-)$
2, 15	Right $(+/-)$
5, 18	Bit clock $(+/-)$
6, 19	WCLK $(+/-)$
7, 20	Master clock $(+/-)$
12, 25	GND

a separate bit clock (1.152 MHz square wave at 48 kHz sampling rate) and a word clock signal taking the same form as the multitrack version. No record status information was carried over this interface but an additional 'master clock' was offered at 2.304 MHz. The pin assignments are shown in Table 4.13.

4.12.8 Sony to Mitsubishi conversion

Comparing the SDIF format with the Mitsubishi multichannel format it may be appreciated that to interconnect the two would only require minor modifications to the signals. Both have a 32-bit structure, MSB first, with only 16 bits being used for audio resolution, the difference being the sync pattern in Sony's bits 29–32, plus the fact that the Mitsubishi does not include control and user bits. The word clock of the Sony is a square wave at the sampling frequency, whereas that of Mitsubishi only goes low for one bit period, but simple logic could convert from one to the other. If transferring from Sony to Mitsubishi it would be necessary also to derive a bit clock, and this could be multiplied up from the word clock using a suitable phase-locked loop. Commercial interfaces are available which perform this task neatly.

4.12.9 Yamaha interface

Yamaha digital audio equipment is often equipped with a 'cascade' connector to allow a number of devices to be operated in cascade, such that the two-channel mix outputs of a mixer, for example, may be fed into a further mixer to be combined with another mixed group of channels. The so-called Y1 format is monophonic and the Y2 format carries two channels in serial form. The two types are very similar, Y1 simply ignoring the second channel of data.

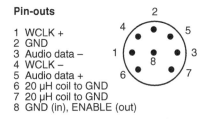

Figure 4.35 Pinouts of the Yamaha two-channel cascade interface.

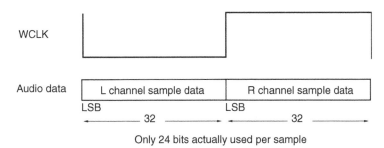

Figure 4.36 Data format of the Yamaha 'cascade' interface.

The two-channel cascade interface terminates in an eight pin DIN-type connector, as shown in Figure 4.35, and carries two channels of 24-bit audio data over an RS-422-standard differential line. The two channels of data are multiplexed over a single serial link, with a 32-bit word of left channel data followed by a 32-bit word of right channel data (the 24 bits of audio are sent LSB first, followed by eight zeros). The word clock alternates between low state for the left channel and high state for the right channel, as shown in Figure 4.36. Coils of 20 μH are connected between pins 6 and 7 and ground to enable suppression of radio frequency interference. The OUT socket is only enabled when its pin 8 is connected to ground.

References

1. AES, *AES3-1985 (ANSI S4.40-1985). Serial transmission format for linearly represented digital audio data. Journal of the Audio Engineering Society*, vol. 33, pp. 975–984 (1985)
2. Finger, R., AES3-1992: the revised two channel digital audio interface. *Journal of the Audio Engineering Society*, vol. 40, March, pp. 107–116 (1992)
3. EBU, *Tech. 3250-E. Specification of the digital audio interface.* Technical Centre of the European Broadcasting Union, Brussels (1985)
4. CCIR, *Rec. 647. A digital audio interface for broadcasting studios.* Green Book, vol. 10, pt. 1. International Radio Consultative Committee, Geneva (1986)

5. CCIR, *Rec. 647 (Mod. F). Draft digital audio interface for broadcast studios.* CCIR, Geneva (1990)

6. IEC, *IEC 958. Digital audio interface, first edition.* International Electrotechnical Commission, Geneva (1989)

7. EIAJ, *CP-340. A digital audio interface.* Electronic Industries Association of Japan, Tokyo (1987)

8. EIAJ, *CP-1201. Digital audio interface (revised).* Electronic Industries Association of Japan, Tokyo (1992)

9. British Standards Institute, *BS 7239. Specification for digital audio interface.* British Standards Institute, London (1989)

10. IEC, *IEC 60958-1 to 4. Digital audio interface, parts 1–4.* International Electrotechnical Commission, Geneva (1999)

11. AES, *AES3-1992 (r1997) (ANSI S4.40-1992). Serial transmission format for linearly represented digital audio data* (1992)

12 AES, *AES10-1991 (ANSI S4.43-1991).* Serial multichannel audio digital interface (MADI). *Journal of the Audio Engineering Society,* vol. 39, pp. 369–377 (1991)

13. CCITT, *Rec. V.11. Electrical characteristics for balanced double-current interchange circuits for general use with integrated circuit equipment in the field of data communications.* International Telegraph and Telephone Consultative Committee (1976, 1980)

14. EIA, *Industrial electronics bulletin no. 12.* EIA standard RS-422A. Electronics Industries Association, Engineering Dept., Washington, DC

15. Dunn, J., Considerations for interfacing digital audio equipment to the standards AES3, AES5 and AES11. In *Proceedings of the AES 10th International Conference,* 7–9 September, p. 122, Audio Engineering Society (1991)

16. Ajemian, R.G. and Grundy, A.B., Fiber optics – the new medium for audio: a tutorial. *Journal of the Audio Engineering Society,* vol. 38, March, pp. 160–175 (1990)

17. AES, *AES-3-ID. AES information document for digital audio engineering – transmission of AES3 formatted data by unbalanced coaxial cable* (2001)

18. SMPTE, *SMPTE 276M: AES/EBU audio over coaxial cable* (1995)

19. Rorden, B. and Graham, M., A proposal for integrating digital audio distribution into TV production. *JSMPTE,* September, pp. 606–608 (1992)

20. AES, *AES-2-ID. AES information document for digital audio engineering – guidelines for the use of the AES3 interface* (1996)

21. Gilchrist, N., Coordination signals in the professional digital audio interface. In *Proceedings of the AES/EBU Interface Conference,* 12–13 September, pp. 13–15, Audio Engineering Society British Section (1989)

22. Komly, A. and Viallevieille, A., Programme labelling in the user channel. In *Proceedings of the AES/EBU Interface Conference,* 12–13 September, pp. 28–51, Audio Engineering Society British Section (1989)

23. ISO 3309, *Information processing systems – data communications – high level data link frame structure.* International Organization for Standardization (1984)

24. AES18, Format for the user data channel of the AES digital audio interface. *Journal of the Audio Engineering Society,* vol. 40, no. 3, March, pp. 167–183 (1992)

25. Nunn, J.P., Ancillary data in the AES/EBU digital audio interface. In *Proceedings of the 1st NAB Radio Montreux Symposium,* 10–13 June, pp. 29–41 (1992)

26. AES, *AES18-1996. Format for the user data channel of the AES digital audio interface* (1996)

27. DAT Conference Part V, *Digital audio taperecorder system (RDAT). Recommended design standard* (1986)

28. EBU, *EBU Technical Recommendation R72-1999: Allocation of the audio modes in the digital audio interface (EBU document Tech. 3250)* (1999)

29. Rumsey, F., *Spatial Audio.* Focal Press (2001)

30. Sanchez, An understanding and implementation of the SCMS serial copy management system for digital audio transmission. *Journal of the Audio Engineering Society,* vol. 42, no. 3, March, pp. 162–186 (1994)

31. IEC, *IEC 61937 Digital audio – interface for non-linear PCM encoded audio bitstreams applying IEC 60958* (2000)
32. SMPTE, *SMPTE 337M: Television – format for non-PCM audio and data in AES3 serial digital audio interface* (2000)
33. AES, *AES42-2001: AES standard for acoustics – digital interface for microphones* (2001)
34. Wilton, P., *MADI (Multichannel audio digital interface)*. In *Proceedings of the AES/EBU Interface Conference*, 12–13 September, pp. 117–130, Audio Engineering Society British Section (1989)
35. Rumsey, F.J., *MIDI Systems and Control*, 2nd ed. Focal Press (1994)
36. Page, M. *et al.*, Multichannel audio connection for Direct Stream Digital. Presented at AES 113th Convention, Los Angeles, 5–8 October (2002)

5

Carrying real-time audio over computer interfaces

There is an increasing trend towards employing standard computer interfaces and networks to transfer audio information, as opposed to dedicated audio interfaces. Such computer interfaces are typically used for a variety of purposes in general data communications and they may need to be adapted for audio applications that require sample-accurate real-time transfer. The increasing ubiquity of computer systems in audio environments makes it inevitable that generic data communication technology will gradually take the place of dedicated interfaces. It also makes sense economically to take advantage of the 'mass market' features of the computer industry.

The applications and protocols described in this chapter are primarily concerned with real-time audio communications or 'streaming', rather than file transfer, and the coverage is limited to studio contexts. The reason for this is that these applications are similar to those for which the dedicated audio interfaces described elsewhere in this book would be used. Some examples are given of proprietary technology that addresses the problems of streaming audio over computer networks but not every proprietary solution is covered in detail. The wider issue of Internet audio streaming for the consumer distribution of music or broadcasts is not covered in any detail. Those interested in more detailed aspects of network applications in digital audio are referred to Andy Bailey's book *Network Technology for Digital Audio*[1].

5.1 Introduction to carrying audio over computer interfaces

Dedicated audio interfaces mostly carry audio data in a sample-clock-synchronized fashion, often with an embedded sample clock signal, carrying little or no other data than that required to move audio samples for one or more channels between devices. They behave like 'digital wires', acting as the digital equivalent of an analog signal cable, connecting one point in a system directly to another. Computer interfaces, on the other hand, are typically general-purpose data carriers that may have asynchronous features and may not always have the inherent quality-of-service features that are required for 'streaming' applications. They also normally use an addressing structure that enables packets of data to be carried from one of a number of sources to one of a number of destinations and such packets will share the connection in a more or less controlled way. Data transport protocols such as TCP/IP are often used as a universal means of managing the transfer of data from place to place, adding overheads in terms of data rate, delay and error handling that may work against the efficient transfer of audio. Data interfaces may be intended primarily for file transfer applications where the time taken to transfer the file is not a crucial factor – as fast as possible will do.

Conventional office Ethernet is a good example of a computer network interface that has limitations in respect of audio streaming. The original 10 Mbit/s data rate was quite slow, although theoretically capable of handling a number of channels of real-time audio data. If employed between only two devices and used with a low-level protocol such as UDP (user datagram protocol) audio can be streamed quite successfully, but problems can arise when multiple devices contend for use of the bus and where the network is used for general-purpose data communications in addition to audio streaming. There is no guarantee of a certain quality of service, because the bus is a sort of 'free for all', 'first-come-first-served' arrangement that is not designed for real-time applications. To take a simple example, if one's colleague attempts to download a huge file from the Internet just when one is trying to stream a broadcast live to air in a local radio station, using the same data network, the chances are that one's broadcast will drop out occasionally.

One can partially address such limitations in a crude way by throwing data-handling capacity at the problem, hoping that increasing the network speed to 100 Mbit/s or even 1 Gbit/s will avoid it ever becoming overloaded. Circuit-switched networks can also be employed to ease these problems (that is networks where individual circuits are specifically established between sources and destinations). Unless capacity can be reserved and service quality guaranteed a data network will never be a suitable replacement for dedicated audio interfaces in critical environments such as broadcasting stations. This has led to the development of real-time protocols and/or circuit-switched networks for handling audio information on data interfaces, in which latency (delay) and bandwidth are defined and guaranteed. The audio industry can benefit from the increased data rates, flexibility and versatility of general-purpose interfaces provided that these issues are taken seriously.

Desktop computers and consumer equipment are also increasingly equipped with general-purpose serial data interfaces such as USB (universal serial bus) and FireWire (IEEE 1394). These have a high enough data rate to carry a number of channels of audio data over relatively short distances, either over copper or optical fibre. Audio protocols also exist for these, as described below.

5.2 Audio over FireWire (IEEE 1394)

5.2.1 Basic FireWire principles

FireWire is an international standard serial data interface specified in IEEE 1394–1995[2]. One of its key applications has been as a replacement for SCSI (Small Computer Systems Interface) for connecting disk drives and other peripherals to computers. It is extremely fast, running at rates of 100, 200 and 400 Mbit/s in its original form, with higher rates appearing all the time up to 3.2 Gbit/s. It is intended for optical fibre or copper interconnection, the copper 100 Mbit/s (S100) version being limited to 4.5 m between hops (a hop is the distance between two adjacent devices). The S100 version has a maximum realistic data capacity of 65 Mbit/s, a maximum of 16 hops between nodes and no more than 63 nodes on up to 1024 separate buses. On the copper version there are three twisted pairs – data, strobe and power – and the interface operates in half duplex mode, which means that communications in two directions are possible, but only one direction at a time. The 'direction' is determined by the current transmitter which will have been arbitrated for access to the bus. Connections are 'hot pluggable' with auto-reconfiguration – in other words one can connect and disconnect devices without turning off the power and the remaining system will reconfigure itself accordingly. It is also relatively cheap to implement.

Unlike, for example, the AES3 audio interface, data and clock (strobe) signals are separated. A clock signal can be derived by exclusive-OR'ing the data and strobe signals, as shown in Figure 5.1. FireWire combines features of network and point-to-point interfaces, offering both asynchronous and isochronous communication modes, so guaranteed latency and bandwidth are available if needed for time-critical applications. Communications are established between logical addresses, and the end point of an isochronous stream is called a 'plug'. Logical connections between devices can be specified as either 'broadcast' or 'point-to-point'. In the broadcast case either the transmitting or receiving plug is defined, but not both, and broadcast connections are unprotected in that any device can start and stop it. A primary advantage for audio applications is that point-to-point connections are protected – only the device that initiated a transfer can interfere with that connection, so once established the data rate is guaranteed for as long as the link remains intact. The interface can be used for real-time multichannel audio interconnections, file transfer, MIDI and machine control, carrying digital video, carrying any other computer data and connecting peripherals (e.g. disk drives).

Data is transferred in packets within a cycle of defined time (125 μs) as shown in Figure 5.2. The data is divided into 32-bit 'quadlets' and

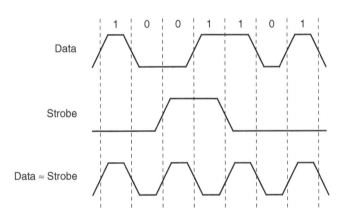

Figure 5.1 Data and strobe signals on the 1394 interface can be exclusive-OR'ed to create a clock signal.

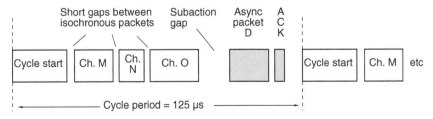

Figure 5.2 Typical arrangement of isochronous and asynchronous packets within a 1394 cycle.

isochronous packets (which can be time stamped for synchronization purposes) consist of between 1 and 256 quadlets (1024 bytes). Packet headers contain data from a cycle time register that allows for sample accurate timing to be indicated. Resolutions down to about 40 nanoseconds can be indicated. One device on the bus acts as a bus master, initiating each cycle with a cycle start packet. Subsequently devices having isochronous packets to transmit do so, with short gaps between the packets, followed by a longer subaction gap after which any asynchronous information is transmitted.

5.2.2   Audio and Music Data Transmission Protocol

Originating partly in Yamaha's 'm-LAN' protocol, the 1394 Audio and Music Data Transmission Protocol[3] is now also available as an IEC PAS component of the IEC 61883 standard[4] (a PAS is a publically available specification that is not strictly defined as a standard but is made available for information purposes by organizations operating under given procedures). It offers a versatile means of transporting digital audio and MIDI control data. It specifies that devices operating this protocol should be capable of the 'arbitrated short bus reset' function which ensures that audio transfers are not interrupted during bus resets. Those wishing to implement this protocol should, of course, refer directly to the standard, but a short summary of some of the salient points is given here.

The complete model for packetizing audio data so that it can be transported over the 1394 interface is complex and very hard to understand, but some applications make the overall structure seem more transparent, particularly if the audio samples are carried in a simple 'AM824' format, each quadlet of which has an eight-bit label and 24 bits of data. The model is layered as shown in Figure 5.3 in such a way that audio applications generate data that is formed (adapted) into blocks or clusters with appropriate labels and control information such as information about the nominal sampling frequency, channel configuration and so forth. Each block contains the information that arrives for transmission within one audio sample period, so in a surround sound application it could be a sample of data for each of six channels of audio plus related control information. The blocks, each representing 'events', are then 'packetized' for transmission over the interface. The so-called 'CIP layer' is the common isochronous packet layer that is the transport stream of 1394. Each isochronous packet has a header that is two quadlets long, defining it as an isochronous packet and indicating its length, and a two quadlet CIP header that describes the following data as audio/music data and indicates (among

Example

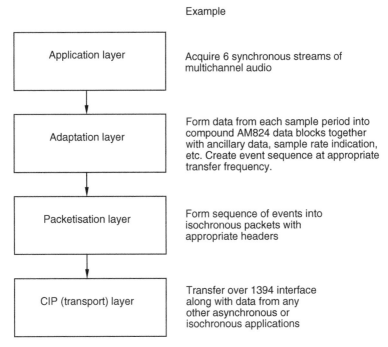

Application layer — Acquire 6 synchronous streams of multichannel audio

Adaptation layer — Form data from each sample period into compound AM824 data blocks together with ancillary data, sample rate indication, etc. Create event sequence at appropriate transfer frequency.

Packetisation layer — Form sequence of events into isochronous packets with appropriate headers

CIP (transport) layer — Transfer over 1394 interface along with data from any other asynchronous or isochronous applications

Figure 5.3 Example of layered model of 1394 audio/music protocol transfer.

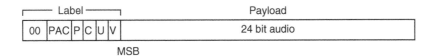

PAC = preamble code (takes place of preamble sync pattern in conventional digital interface)
11 = Z (or B)
01 = X (or M)
00 = Y (or W)

Figure 5.4 AM824 data structure for IEC 60958 audio data on 1394 interface. Other AM824 data types use a similar structure but the label values are different to that shown here.

other things) the presentation time of the event for synchronization purposes. A packet can contain more than one audio event and this becomes obvious when one notices that the cycle time of 1394 (the time between consecutive periods in which a packet can be transmitted) is normally 125 μs and an audio sample period at 48 kHz is only 22 μs.

1394 can carry audio data in IEC 60958 format (see section 4.3). This is based on the AM824 data structure in which the eight-bit label serves as a substitute for the preamble and VUCP data of the IEC subframe, as shown in Figure 5.4. The following 24 bits of data are then simply the audio data component of the IEC subframe. The two subframes forming an IEC frame are transmitted within the same event and each has to have the eight-bit label at the start of the relevant quadlet (indicating left or right channel).

The same AM824 structure can be used for carrying other forms of audio data including multibit linear audio (a raw audio data format used in some DVD applications, termed MBLA), high resolution MBLA, one-bit audio (see section 4.12.2), MIDI, SMPTE timecode and sample count or ancillary data. These are indicated by different eight-bit labels. One-bit audio can be either raw or DST (Direct Stream Transfer) encoded. DST is a lossless data reduction system employed in Direct Stream Digital equipment and Super Audio CD.

Audio data quadlets in these different modes can be clustered into compound data blocks. As a rule a compound data block contains samples from a number of related streams of audio and ancillary information that are based on the same sampling frequency table (see next section). The parts of these blocks can be application specific or unspecific. In general, compound blocks begin with an unspecified region (although this is not mandatory) followed by one or more application-specific regions (see Figure 5.5). The unspecified region can contain audio/music content data and it is recommended that this always starts with basic two-channel stereo data in either IEC or raw audio format, followed by any other

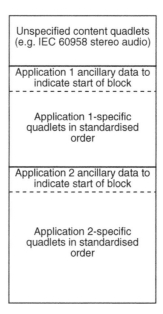

Figure 5.5 General structure of a compound data block.

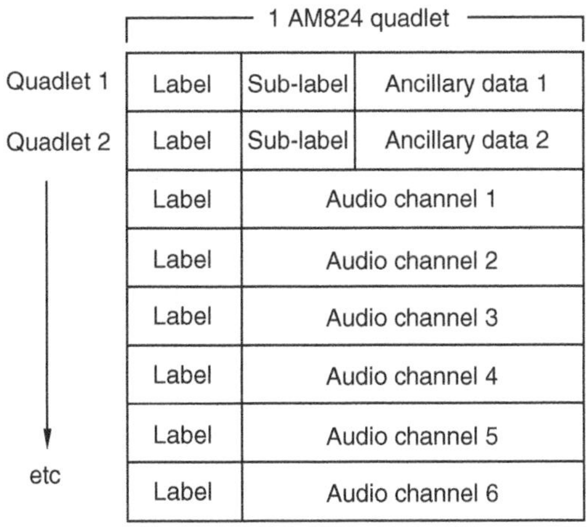

Figure 5.6 Specific example of an application-specific data block for multichannel audio transfer from a DVD player.

unspecified content data in a recommended order. An example of an application-specific part is the transfer of multiple synchronous channels from a DVD player. Here ancillary data quadlets indicate the starts of blocks and control factors such as downmix values, multichannel type (e.g. different surround modes), dynamic range control and channel assignment. An example of such a multichannel cluster is shown in Figure 5.6.

5.2.3 Clock synchronization

It is also possible to transfer information relating to the synchronization of a sample clock using the audio/music protocol over 1394. The instantaneous *actual* sampling frequency (the rate at which the audio system is running) can be worked out from the time stamps contained in the SYT part of packet headers and the SYT interval (the number of data blocks or sample periods between two successive time stamps). Audio clock information can be derived from this and a receiver clock can be controlled. The SYT time stamp is intended to indicate the time at which the event concerned is to be presented to the receiver and is not supposed to be transmitted at a rate lower than 3.5 kHz under normal circumstances. If there is more than one event in a packet then this usually corresponds to the start of the first one. Professional receivers are supposed to make it possible to use the SYT information to control the presentation time of events, but consumer devices where implementation costs are critical do not have to do this and can 'free run'.

The nominal sampling frequency is usually indicated as a part of the CIP header, this being a formal means of indicating to a receiver what the intended sampling frequency should be (rather like the sampling frequency indicated in channel status of AES3). It is contained within the three LSBs of the format dependent field (FDF) of the CIP header and there are a number of defined tables that show the relationship between values of the sample frequency code (SFC) and the corresponding SYT interval for different transmission modes.

5.3 Audio over universal serial bus (USB)

5.3.1 Basic USB principles

The universal serial bus is not the same as IEEE1394, but it has some similar implications for desktop multimedia systems, including audio peripherals. USB has been jointly supported by a number of manufacturers including Microsoft, Digital, IBM, NEC, Intel and Compaq. It is a copper interface that, in its basic version, runs at a lower speed than 1394 (typically either 1.5 or 12 Mbit/s) and is designed to act as a low cost connection for multiple input devices to computers such as joysticks, keyboards, scanners and so on. The data rate is, however, high enough for it to be used for transferring limited audio information if required. A recent revision of the USB standard enables newer interfaces to operate at a high rate of up to 480 Mbit/s.

USB supports up to 127 devices for both isochronous and asynchronous communication and can carry data over distances of up to 5 m per hop (similar to 1394). A hub structure is required for multiple connections to the host connector. Like 1394 it is hot pluggable and reconfigures the addressing structure automatically. When new devices are connected to a USB setup the host device assigns a unique address. Limited power is available over the interface and some devices are capable of being powered solely using this source – known as 'bus-powered' devices – which can be useful for field operation of, say, a simple A/D convertor with a laptop computer.

Data transmissions are grouped into frames of 1 ms duration in USB 1.0 but a 'micro-frame' of one-eighth of 1 ms was also defined in USB 2.0. A start-of-frame packet indicates the beginning of a cycle and the bus clock is normally at 1 kHz if such packets are transmitted every millisecond. So the USB frame rate is substantially slower than the typical audio sampling rate. The transport structure and different layers of the network protocol will not be described in detail as they are long and complex and can be found in the USB 2.0 specification[5]. However, it is important to be aware that transactions are set up between sources and destinations over so-called 'pipes' and that numerous 'interfaces' can be defined and run over a single USB cable, only dependent on the available bandwidth. Some salient features of the audio specification will be described.

5.3.2 Audio over USB

The way in which audio is handled on USB is well defined and somewhat more clearly explained than the 1394 audio/music protocol[6]. It defines three types of communication: audio control, audio streaming and MIDI streaming. We are concerned primarily with audio streaming applications.

Audio data transmissions fall into one of three types. Type 1 transmissions consist of channel-ordered PCM samples in consecutive subframes, whilst Type 2 transmissions typically contain non-PCM audio data that does not preserve a particular channel order in the bitstream, such as certain types of multichannel data-reduced audio stream. Type 3 transmissions are a hybrid of the two such that non-PCM data is packed into pseudo-stereo data words in order that clock recovery can be made easier. This method is in fact very much the same as the way data-reduced audio is packed into audio subframes within the IEC 61937 format described in Chapter 4, and follows much the same rules.

Audio samples are transferred in subframes, each of which can be one to four bytes long (up to 24 bits resolution). An audio frame consists of one or more subframes, each of which represents a sample of different channel in the cluster (see below). As with 1394, a USB packet can contain a number of frames in succession, each containing a cluster of subframes. Frames are described by a format descriptor header that contains a number of bytes describing the audio data type, number of channels, subframe size, as well as information about the sampling frequency and the way it is controlled (for Type 1 data). An example of a simple audio frame would be one containing only two subframes of 24-bit resolution for stereo audio.

Audio of a number of different types can be transferred in Type 1 transmissions, including PCM audio (two's complement, fixed point), PCM-8 format (compatible with original eight-bit WAV, unsigned, fixed point), IEEE floating point, A-law and μ-law (companded audio corresponding to relatively old telephony standards). Type 2 transmissions typically contain data-reduced audio signals such as MPEG or AC-3 streams. Here the data stream contains an encoded representation of a number of channels of audio, formed into encoded audio frames that relate to a large number of original audio samples. An MPEG encoded frame, for example, will typically be longer than a USB packet (a typical MPEG frame might be 8 or 24 ms long), so it is broken up into smaller packets for transmission over USB rather like the way it is streamed over the IEC 60958 interface described in Chapter 4. The primary rule is that no USB packet should contain data for more than one encoded audio frame, so a new encoded frame should always be started in a new packet. The format descriptor for Type 2 is similar to Type 1 except that it replaces subframe size and number of channels indication with maximum bit rate and number of audio samples per encoded frame. Currently only MPEG and AC-3 audio are defined for Type 2.

Rather like the compound data blocks possible in 1394 (see above), audio data for closely related synchronous channels can be clustered for USB transmission in Type 1 format. Up to 254 streams can be clustered and there are 12 defined spatial positions for reproduction, to simplify the relationship between channels and the loudspeaker locations to which they relate. (This is something of a simplification of the potentially complicated formatting of spatial audio signals and assumes that channels are tied to loudspeaker locations, but it is potentially useful.) The first six defined streams follow the internationally standardized order of surround sound channels for 5.1 surround, that is left, right, centre, LFE (low frequency effects), left surround, right surround. Subsequent streams are allocated to other loudspeaker locations around a notional listener. Not all the spatial location streams have to be present but they are supposed to be presented in the defined order. Clusters are defined in a descriptor field that includes 'bNrChannels' (specifying how many logical audio channels are present in the cluster) and 'wChannelConfig' (a bit field that indicates which spatial locations are present in the cluster). If the relevant bit is set then the relevant location is present in the cluster. The bit allocations are shown in Table 5.1.

5.3.3 Clock synchronization

Audio devices transferring signals over USB can have sample clocks that are either asynchronous with the USB data transfer, or that are locked in some way to the USB start-of-frame (SOF) identifier (that occurs every 1 ms). Asynchronous devices would typically use free-running or externally synchronized audio clocks, whereas synchronous devices would either have a means of locking their sample clocks to the 1 ms SOF point

Table 5.1 Channel identification in USB audio cluster descriptor

Data bit	*Spatial location*
D0	Left Front (L)
D1	Right Front (R)
D2	Center Front (C)
D3	Low Frequency Enhancement (LFE)
D4	Left Surround (Ls)
D5	Right Surround (Rs)
D6	Left of Center (Lc)
D7	Right of Center (Rc)
D8	Surround (S)
D9	Side Left (Sl)
D10	Side Right (Sr)
D11	Top (T)
D15..12	Reserved

or (perhaps unusually) have a means of controlling the USB clock rate so that it became locked to the audio sampling frequency. It is up to host applications to ensure that groups of audio channels that belong together and are supposed to be sample-aligned are kept so through any buffering that is employed. Buffering of at least one frame is normally required at the receiver and the management and reporting of delays is an inherent feature of the recommendations.

5.4   AES47: audio over ATM

Asynchronous transfer mode (ATM) is a protocol for data transmission over high speed data networks that operates in a switched fashion and can extend over wide or metropolitan areas. It typically operates over SONET (synchronous optical network) or SDH (synchronous digital hierarchy) networks, depending on the region of the world. Switched networks involve the setting up of specific connections between a transmitter and one or more receivers, rather like a dialled telephone network (indeed this is the infrastructure of the digital telephone network). Data packets on ATM networks consist of a fixed 48 bits, typically preceded by a five-byte header that identifies the virtual channel of the packet.

AES47[7] defines a method by which linear PCM data, either conforming to AES3 format or not, can be transferred over ATM. There are various arguments for doing this, not least being the increasing use of such networks for data communications within the broadcasting industry and the need to route audio signals over longer distances than possible using standard digital interfaces. There is also a need for low latency, guaranteed bandwidth and switched circuits, all of which are features of ATM. Essentially an ATM connection is established in a similar way to making a telephone call. A SETUP message is sent at the start of a new 'call' that describes the nature of the data to be transmitted and defines its vital statistics. The AES47 standard describes a specific professional audio implementation of this procedure that includes information about the audio signal and the structure of audio frames in the SETUP at the beginning of the call.

For some reason bytes are termed octets in ATM terminology, so this section will follow that convention. Audio data is divided into subframes and each subframe contains a sample of audio as well as optional ancillary data and protocol overhead data, as shown in Figure 5.7. The setup message at the start of the call determines the audio mode and whether or not this additional data is present. The subframe should occupy a whole number of octets and the length of the audio sample should be such that the subframe is 8, 16, 24, 32 or 48 bits long. The ancillary data field, if it is present, is normally used for carrying the VUC bits from the AES3 subframe, along with a B bit to replace the P (parity) bit of the AES3 subframe (which has little relevance in this new application). The B bit in the '1' state indicates the start of an AES3 channel status block, taking the place of the Z preamble that is no longer present. This data is transmitted in the order BCUV.

The protocol overhead bits, if present, consist of a sequencing bit followed by three data protection bits (used for error checking). These sequencing bits are assembled from all the subframes in an ATM cell, rather as channel status bits are assembled from successive AES3 subframes to form a sequencing word. The first four bits of this form the sequencing number, the point of which is to act as an incremented count of ATM cells since the start of the call. Bits 5–7 act as protection bits for the sequencing word, bit 8 is even parity for the first eight bits, and bits 9–12 (if present) can form a second sequencing number that can be used to align samples from multiple virtual circuits carrying nominally time-aligned signals (see Figure 5.8).

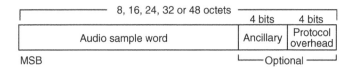

Figure 5.7   General audio subframe format of AES47.

First subframe's sequencing bit

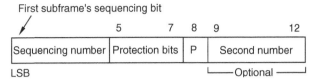

Figure 5.8 Components of the sequencing word in AES47.

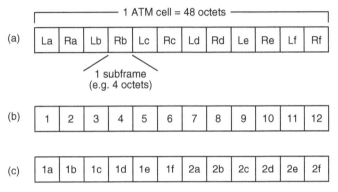

Figure 5.9 Packing of audio subframes into ATM cells. (a) Example of temporal ordering with two channels, left and right. 'a', 'b', 'c', etc., are successive samples in time for each channel. Co-temporal samples are grouped together. (b) Example of multichannel packing whereby concurrent samples from a number of channels are arranged sequentially. (c) Example of ordering by channel, with a number of samples from the same channel being grouped together. (If the number of channels is the same as the number of samples per cell, all three methods turn out to be identical.)

Samples are packed into the ATM cell either ordered in time, in multichannel groups or by channel, as shown in Figure 5.9. Only certain combinations of channels and data formats are allowed and all the channels within the stream have to have the same resolution and sampling frequency, as shown in Table 5.2.

Four octets in the user-defined AAL part of the SETUP message that begins a new ATM call define aspects of the audio communication that will take place. The first byte contains so-called 'qualifying information', only bit 4 of which is currently specified indicating that the sampling frequency is locked to some global reference. The second byte indicates the subframe format and sample length, whilst the third byte specifies the packing format. The fourth byte contains information about the audio sampling frequency (32, 44.1 or 48 kHz), its scaling factor (from 0.25 up to 8 times) and multiplication factor (e.g. 1/1.001 or 1.001/1 for 'pull-down' or 'pull-up' modes). It also has limited information for varispeed rates.

Table 5.2 Audio packing within ATM cells – options in AES47

AAL code (hex)*	Subframe length (bytes)	Audio resolution	Ancillary bits	Protocol bits	Grouping	No. of audio channels
56 02	4	24	4	4	Temporal	2
56 01	4	24	4	4	N/A	1
06 02	3	24	0	0	Temporal	2
06 01	3	24	0	0	N/A	1
56 85	4	24	4	4	Multichannel	60

* This should be signalled within the second and third octets of the user-defined AAL part of the SETUP message that is an optional part of the ATM protocol for setting up calls between sources and destinations.

There is provision within the standard for the sender to include a local clock that ticks once per second. It is expected that cells will be blocked such that a block consists of either eight cells or eight sets of samples. The User Indication (UI) bit in the cell header should be set to 1 in the first and last cells of the first block following a clock tick. This can be used to derive a pulse train related to the sampling frequency.

5.5 ISDN

ISDN, the Integrated Services Digital Network, is an extension of the digital telephone network to the consumer, providing two 64 Kb/s digital channels that can be connected to ISDN terminals anywhere in the world by dialling. Since the total usable capacity of an ISDN 'B' connection is only 128 Kb/s it is not possible to carry linear PCM data at normal audio resolutions over such a link, but it is possible to carry moderately high quality stereo audio at this rate using a data reduction system such as MPEG Layer 3[8], or to achieve higher rates by combining more than one ISDN link to obtain data rates of, say, 256 or 384 Kb/s[9]. 'Broadband ISDN', on the other hand, is capable of much higher data rates than a simple 'B' channel connection and is in fact another name for ATM networking, described in the previous section.

5.6 CobraNet

CobraNet is a proprietary audio networking technology developed by Peak Audio, a division of Cirrus Logic. It is designed for carrying audio over conventional Fast Ethernet networks (typically 100 Mbit/s), preferably using a dedicated Ethernet for audio purposes or using a switched Ethernet network. Switched Ethernet acts more like a telephone or ATM network where connections are established between specific sources and destinations, with no other data sharing that 'pipe'. For the reasons stated earlier in this chapter, Ethernet is not ideal for audio communications without some provisos being observed. CobraNet, however, implements a method of arbitration, bandwidth reservation and an isochronous transport protocol that enables it to be used successfully.

It is claimed that conventional data communications and CobraNet applications can coexist on the same physical network but the system implements new arbitration rules, in the form of a so-called 'O-persistent' layer within the data link layer, to ensure that collisions do not take place on the network. All devices have to be able to abide by these rules if they are to be used. The company provides software that can be used to verify that an existing network design is capable of handling the audio information intended (in respect of bandwidth, delay and other critical parameters).

CobraNet can also be used for sample clock distribution and for equipment control purposes (interfacing with RS-232 and RS-485 equipment) and it is becoming popular in venue or live sound installations for transferring audio between multiple locations. It requires a dedicated CobraNet interface to convert audio and control data streams to and from the relevant Ethernet protocol and claims a low audio latency of 5.3 ms. A number of items of audio equipment are already equipped with CobraNet technology, such as microphone preamplifiers, power amplifiers and routers. Users claim considerable benefits in being able to integrate audio transport and equipment control/monitoring using a single network interface.

The CobraNet protocol has been allocated its own protocol identifier at the Data Link Layer of the ISO 7-layer network model, so it does not use Internet Protocol (IP) for data transport (this is typically inefficient for audio streaming purposes and involves too much overhead). Because it does not use IP it is not particularly suitable for wide area network (WAN) operation and would typically be operated over a local area network (LAN). It does, however, enable devices to be allocated IP addresses using the BOOTP (boot protocol) process and supports the use of IP and UDP (user datagram protocol) for other purposes than the carrying of audio. It is capable of transmitting packets in isochronous cycles, each packet transferring data for a 'bundle' of audio channels. Each bundle contains between zero and eight audio channels and these can either be unicast or multicast. Unicast bundles are intended for a single destination whereas multicast bundles are intended for 'broadcast' transmissions whereby a sending device broadcasts packets no matter whether any receiving device is contracted to receive them.

Rather like 1394 (see above) isochronous cycles are initiated by one bus-controlling device (the 'conductor') that sends a multicast packet to indicate the start of a cycle to all other devices ('performers'). In CobraNet terminology this is called the 'beat packet'. This beat packet is a form of clock for the network and also carries information about the overall network operation, so it is sensitive to delays and must be maintained within a very narrow time window ($250\,\mu s$) if it is to be used for sample clock locking. These packets are typically small (100 bytes) whereas audio data packets are typically much larger (e.g. 1000 bytes). The CobraNet interface derives a sample clock from the network clock by means of a VCXO (voltage-controlled crystal oscillator) circuit.

5.7 MAGIC

MAGIC[10] (Media-accelerated Global Information Carrier) was developed by the Gibson Guitar Corporation, originally going under the name GMICS. It is a relatively recent audio interface that typically uses the Ethernet physical layer for transporting audio between devices, although it is not compatible with higher layers and does not appear to be interoperable with conventional Ethernet data networks. It uses its own application and data link layers, the data link layer of which is based on the Ethernet 802.3 data link layer, using a frame header that would be recognized by 802.3-compatible devices.

Although it is not limited to doing so, the described implementation uses 100 Mbit/s Fast Ethernet over standard CAT 5 cables, using four of the wires in a conventional Ethernet crossover implementation and the other four for power to devices capable of operating on limited power (9 volt, 500 mA). Data is formed into frames of 55 bytes, including relevant headers, and transmitted at a synchronous rate between devices. The frame rate is related to the audio sampling rate and a sampling clock can be recovered from the interface. Very low latency of $10-40\,\mu s$ is claimed. MAGIC data can be daisy-chained between devices in a form more akin to point-to-point audio interfacing than computer networking, although routing and switching configurations are also possible using routing or switching hubs.

5.8 MOST

MOST (media-oriented synchronous transfer) is described by Heck *et al.*[11] as an alternative network protocol designed for synchronous, asynchronous and control data over a low-cost optical fibre network. It is claimed that the technology sits between USB and IEEE 1394 in terms of performance and that MOST has certain advantages in the transfer of synchronous data produced by multimedia devices that are not well catered for in other protocols. It is stated that interfaces based on copper connections are prone to electromagnetic interference and that the optical fibre interface of this system provides immunity to such, in addition to allowing distances of up to 250 m between nodes in this case.

MOST specifies physical, data link and network layers in the OSI reference model for data networks and dedicated silicon has been developed for the physical layer. Data is transferred in 64-byte frames and the frame rate of data is dependent on the sampling rate in use by the connected devices, being 22.5 Mbit/s at a 44.1 kHz audio sampling rate. The bandwidth can be divided between synchronous and asynchronous data. Potential applications are described including professional audio, for transferring up to 15 stereo 16-bit audio channels or ten stereo channels of 24-bit audio; consumer electronics, as an alternative to SPDIF at similar cost; automotive and home multimedia networking.

There is now a detailed specification framework for MOST[12] and it is the subject of a co-operation agreement between a number of manufacturers. It seems to have been most widely adopted in the automotive industry where it is close to being endorsed by a consortium of car makers.

5.9 BSS SoundWeb

BSS developed its own audio network interface for use with its SoundWeb products that are typically used in large venue installations and for live sound. It uses CAT 5 cabling over distances up to 300 m, but is not based on Ethernet and behaves more like a Token Ring network. Data is carried at a rate of about 12 Mbit/s and transports eight audio channels along with control information.

5.10 Digital content protection

Digital content protection is rather like a more sophisticated version of SCMS, the serial copy management system, described in Chapter 4. Copy protection of digital content is increasingly required by the owners of intellectual property and data encryption is now regarded as the most appropriate way of securing such content from unwanted copying. The

SCMS method used for copy protection on older interfaces such as IEC 60958 involved the use of two bits plus category codes to indicate the copy permission status of content, but no further attempt was made to make the audio content unreadable or to scramble it in the case of non-permitted transfers. A group of manufacturers known as 5C has now defined a method of digital content protection that is initially defined for IEEE 1394 transfers[13] (see section 5.2) but which is likely to be extended to other means of interconnection between equipment. It is written in a relatively generic sense, but the packet header descriptions currently refer directly to 1394 implementations. 5C is the five manufacturers Hitachi, Intel, Matsushita, Sony and Toshiba. The 1394 interface is increasingly used on high-end consumer digital products for content transfer, although it has not been seen much on DVD and SACD players yet because the encryption model has only recently been agreed. There has also been the issue of content watermarking to resolve.

Content protection is managed in this model by means of both embedded copy control information (CCI) and by using two bits in the header of isochronous data packets (the so-called EMI or encryption mode indicator bits). Embedded CCI is that contained within the application-specific data stream itself. In other words it could be the SCMS bits in the channel status of IEC 60958 data or it could be the copy control information in an MPEG transport stream. This can only be accessed once a receiving device has decrypted the data that has been transmitted to it. In order that devices can inspect the copy status of a stream without decrypting the data, the packet header containing the EMI bits is not encrypted. Two EMI bits allow four copy states to be indicated as shown in Table 5.3.

The authentication requirement indicated by the copy state initiates a negotiation between the source and receiver that sets up an encrypted transfer using an exchanged key. The full details of this are beyond the scope of this book and require advanced understanding of cryptography, but it is sufficient to explain that full authentication involves more

Table 5.3 Copy state indication in EMI bits of 1394 header

EMI bit states	Copy state	Authentication required
11	Copy never (Mode A)	Full
10	Copy one generation (Mode B)	Restricted or full
01	No more copies (Mode C)	Restricted or full
00	Copy free(ly) (Mode D)	None (not encrypted)

advanced cryptographic techniques than restricted authentication (which is intended for implementation on equipment with limited or reduced computational resources, or where copy protection is not a major concern). The process is explained in detail in the specification document[13]. The negotiation process, if successful, results in an encrypted and decrypted transfer being possible between the two devices. Embedded CCI can then be accessed from within the content stream.

When there is a conflict between embedded CCI and EMI indications, as there might be during a stream (for example, when different songs on a CD have different CCI but where the EMI setting remains constant throughout the stream) it is recommended that the EMI setting is the most strict of those that will be encountered in the transfer concerned. However, the embedded CCI seems to have the final say-so when it comes to deciding whether the receiving device can record the stream. For example, even if EMI indicates 'copy never', the receiving device can still record it if the embedded CCI indicates that it is recordable. This ensures that a stream is as secure as it should be, and the transfer properly authenticated, before any decisions can be made by the receiving device about specific instances within the stream.

Certain AM824 audio applications (a specific form of 1394 Audio/Music Protocol interchange) have defined relationships between copy states and SCMS states, for easy translation when carrying data like IEC 60958 data over 1394. In this particular case the EMI 'copy never' state is not used and SCMS states are mapped onto the three remaining EMI states. For DVD applications the application-specific CCI is indicated in ancillary data and there is a mapping table specified for various relationships between this data and the indicated copy states. It depends to some extent on the quality of the transmitted data and whether or not it matches that indicated in the *audio_quality* field of ancillary data. (Typically DVD players have allowed single generation home copies of audio material over IEC 60958 interfaces at basic sampling rates – e.g. 48 kHz – but not at very high quality rates such as 96 kHz or 192 kHz.) SuperAudio CD applications currently have only one copy state defined and that is 'no more copies', presumably to avoid anyone duplicating the one-bit stream that would have the same quality as the master recording.

References

1. Bailey, A., *Network Technology for Digital Audio*. Focal Press (2001)
2. IEEE, *IEEE 1394: Standard for a high performance serial bus* (1995)
3. 1394 Trade Association, *TA Document 2001003: Audio and Music Data Transmission Protocol 2.0* (2001)

4. IEC, *IEC/PAS 61883–6. Consumer audio/video equipment – Digital interface – Part 6: Audio and music data transmission protocol* (1998)

5. USB, *Universal serial bus, Revision 2.0 specification.* Available from www.usb.org/developers/docs.html (2000)

6. USB, *Universal serial bus: device class definition for audio devices, v1.0* (1998)

7. AES, *AES47-2002: Transmission of digital audio over asynchronous transfer mode networks* (2002)

8. Brandenburg, K. and Stoll, G., The ISO/MPEG audio codec: a generic standard for coding of high quality digital audio. Presented at the *92nd AES Convention, Vienna, Austria, 24–27 March*, preprint 3336 (1992)

9. Burkhardtsmaier, B. *et al.*, The ISDN MusicTAXI. Presented at the *92nd AES Convention, Vienna, Austria, 24–27 March*, preprint 3344 (1992)

10. Gibson Guitar Corporation, *Media-accelerated Global Information Carrier. Engineering Specification Version 2.4.* Available from www.gibsonmagic.com (2002)

11. Heck, P. *et al.*, Media oriented synchronous transfer: a network protocol for high quality, low cost transfer of synchronous, asynchronous and control data on fiber optic. Presented at the *103rd AES Convention, New York, 26–29 September*, preprint 4551 (1997)

12. Oasis Technology, *MOST Specification Framework v1.1.* Available from: www.oasis.com/technology/index.htm (1999)

13. 5C, *Digital transmission content protection specification, Volume 1 (informational version). Revision 1.2* (2001)

6

Practical audio interfacing

6.1 The importance of synchronization

Unlike analog audio, digital audio has a discrete-time structure, because it is a sampled signal in which the samples may be further grouped into frames and blocks having a certain time duration. If digital audio devices are to communicate with each other, or if digital signals are to be combined in any way, then they need to be synchronized to a common reference in order that the sampling frequencies of the devices are identical and do not drift with relation to each other. It is not enough for two devices to be running at *nominally* the same sampling frequency (say both at 44.1 kHz). Between the sampling clocks of professional audio equipment it is possible for differences in frequency of up to ±10 parts per million (ppm) to exist and even a very slow drift means that two devices are not truly synchronous. Consumer devices can exhibit an even greater range of sampling frequencies that are nominally the same.

The audible effect resulting from a non-synchronous signal drifting with relation to a sync reference or another signal is usually the occurrence of a glitch or click at the difference frequency between the signal and the reference, typically at an audio level around 50 dB below the signal, due to the repetition or dropping of samples. This will appear when attempting to mix two digital audio signals whose sampling rates differ by a small amount, or when attempting to decode a signal such as an unlocked consumer source by a professional system which is locked to a fixed reference. This said, it is not always easy to detect asynchronous operation by listening, even though sample slippage is occurring, as it depends on the nature of audio signal at the time. Some systems may not operate at all if presented with asynchronous signals.

Furthermore, when digital audio is used with analog or digital video, the sampling rate of the audio needs to be locked to the video reference signal and to any timecode signals which may be used. In single studio operations the problem of ensuring lock to a common clock is not as great as it is in a multi-studio centre, or where digital audio signals arrive from remote locations. In distributed system cases either the remote signals must be synchronized to the local sample clock as they arrive, or the remote studio must somehow be fed with the same reference signal as the local studio. A number of approaches may be used to ensure that this happens and they will be explained in this chapter.

Another topic related to synchronization will also be examined here, and that is the importance of short-term clock stability in digitally interfaced audio systems. It is important to distinguish between clock stability requirements in interfacing and clock stability in convertors, although the two are related to some extent.

6.2 Choice of sync reference

6.2.1 AES recommendations

AES recommendations for the synchronization of digital audio signals are documented in AES11-1997[1]. They state that preferably all machines should be able to lock to a reference signal (DARS or 'digital audio reference signal'), which should take the form of a standard two-channel interface signal (see section 4.3) whose sampling frequency is stable within a certain tolerance and that all machines should have a separate input for such a synchronizing signal. If this procedure is not adopted then it is possible for a device to lock to the clock embedded in the channel code of the AES-format audio input signal – a technique known as 'genlock' synchronization. A third option is for the equipment to be synchronized by a master video reference from which a DARS can be derived.

In the AES11 standard signals are considered synchronous when they have identical sampling frequencies, but phase errors are allowed to exist between the reference clock and received/transmitted digital signals in order to allow for effects such as cable propagation delays, phase-locked loop errors and other electrical effects. Input signal frame edges must lie within ±25% of the reference signal's frame edge (taken as the leading edge of the 'X' preamble), and output signals within ±5% (see Figure 6.1), although tighter accuracy than this is preferable because otherwise an

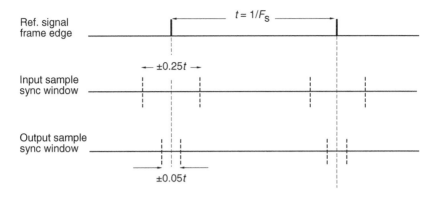

Figure 6.1 Timing limits recommended for synchronous signals in the AES11 standard.

unacceptable build-up of delay may arise when equipment is cascaded. A phase error of $\pm 25\%$ of the frame period is actually quite considerable, corresponding to a timing difference of around $5\,\mu s$ at 48 kHz, so the specification should be readily achievable in most circumstances. For example, a cable length of 58 metres typically gives rise to a delay of only one bit ($0.32\,\mu s$ @ 48 kHz) in the interface signal. If a number of frames' delay exists between the audio input and output of a device, this delay should be stated.

The AES11 reference signal may either contain programme or not. If it does not contain programme it may be digital silence or simply the sync preamble with the rest of the frame inactive. Two grades of reference are specified: Grade 1, having a long-term frequency accuracy of ± 1 ppm (part per million), and Grade 2, having a long-term accuracy of ± 10 ppm. The Grade 2 signal conforms to the standard AES sample frequency recommendation for digital audio equipment (AES5-1984[2]), and is intended for use within a single studio which has no immediate technical reason for greater accuracy, whereas Grade 1 is a tighter specification and is intended for the synchronization of complete studio centres (as well as single studios if required). Byte 4, bits 0 and 1 of the channel status data of the reference signal (see section 4.8) indicate the grade of reference in use (00 = default, 01 = Grade 1, 10 = Grade 2, 11 = reserved). It is also specified in this standard that the capture range of oscillators in devices designed to lock to external synchronizing signals should be ± 2 ppm for Grade 1 and ± 50 ppm for Grade 2. (Grade 1 reference equipment is only expected to lock to other Grade 1 references.)

Despite the introduction of this standard there are still many devices that do not use AES3-format DARS inputs, preferring to rely on BNC-connected word clock, video sync reference signals or other proprietary synchronization clocks.

6.2.2   Other forms of external sync reference

Currently, digital audio recorders are provided with a wide range of sync inputs and most systems may be operated in the external or internal sync modes. In the internal sync mode a system is locked to its own crystal oscillator, which in professional equipment should be accurate within ±10 parts per million (ppm) if it conforms to AES recommended practice, but which in consumer equipment may be much less accurate than this.

In the external sync mode the system should lock to one of its sync inputs, which may either be selectable using a switch, or be selected automatically based on an order of priority depending on the mode of operation (for this the user should refer to the operations manual of the device concerned). Typical sync inputs are word clock (WCLK), which is normally a square-wave TTL-level signal (0–5 V) at the sampling rate, usually available on a BNC-type connector; 'composite video', which is a video reference signal consisting of either normal picture information or just 'black and burst' (a video signal with a blacked-out picture); or a proprietary sync signal such as the optional Alesis sync connection or the LRCK in the Tascam interface (see section 4.12). WCLK may be 'daisy-chained' (looped through) between devices in cases where the AES/EBU interface is not available. Digidesign's ProTools system also uses a so-called 'Super Clock' signal at a multiple of 256 times the sampling rate for slaving devices together with low sampling jitter. This is a TTL level (0–5 V) signal on a BNC connector. In all cases one machine or source must be considered to be the 'master', supplying the sync reference to the whole system, and the others as 'slaves'.

A 'sync' light is usually provided on the front panel of a device (or under a cover) to indicate good lock to an external or internal clock, and this may flash or go out if the system cannot lock to the clock concerned, perhaps because it has too much jitter, is at too low a level, conflicts with another clock, or because it is not at a sampling rate which can be accepted by the system.

Video sync (composite sync) is often used when a system is operated within a video environment, and where a digital recorder is to be referenced to the same sync reference as the video machines in a system. This is useful when synchronizing audio and video transports during recording and replay, using either a synchronizer or a video editor, when timecode is only used initially during the pre-roll, whereafter machines are released to their own video sync reference. It also allows for timecode to be recorded synchronously on the audio machine, since the timecode generator used to stripe the tape can also be locked to video syncs. The relationship between video frame rates and audio sampling frequencies is discussed further in section 6.6.1.

6.3 Distribution of sync references

It is appropriate to consider digital audio as similar to video when approaching the subject of sync distribution, especially in large systems. Consequently, it is advisable to use a central high quality sync signal generator (the equivalent of a video sync pulse generator, or SPG), the output of which is made available widely around the studio centre, using digital distribution amplifiers (DDAs) to supply different outlets in a 'star' configuration. In video operations this is usually called 'house sync'. Each digital device in the system may then be connected to this house sync signal and each should be set to operate in the external sync mode. Because long cable runs or poor quality interfaces can distort digital signals, resulting in timing jitter (see section 6.4.3), it may be advisable to install local reference signal generators slaved to the house sync master generator as a means of providing a 'clean' local reference within a studio. Until such time as AES11 reference inputs become available on audio products one might use word clock or a video reference signal instead.

Using the technique of central sync-signal distribution (see Figure 6.2) it becomes possible to treat all devices as slaves to the sync generator, rather than each one locking to the audio output of the previous one. In this case there is no 'master' machine and the sync generator acts as the 'master'. It requires that all machines in the system operate at the same sampling rate, unless a sampling frequency convertor or signal synchronizer is used (see section 6.5).

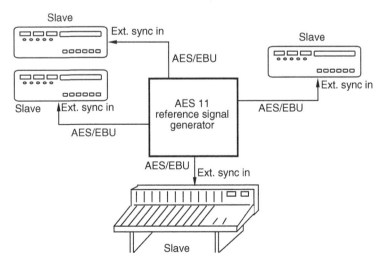

Figure 6.2 All devices within a studio may be synchronized by distributing an AES11 reference signal to external AES sync inputs (if present).

(a)

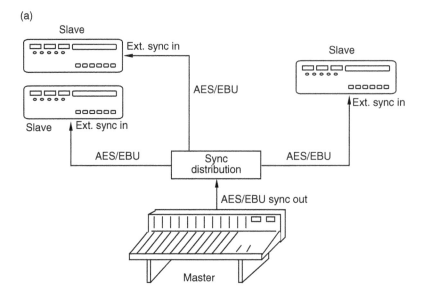

(b)

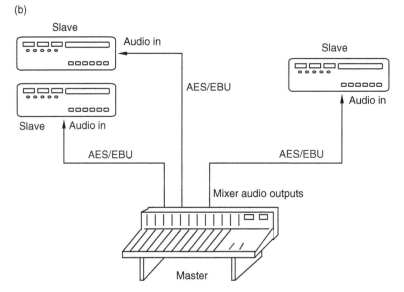

Figure 6.3 (a) In small systems one of the devices, such as a digital mixer, may act as a sync reference for the others; alternatively, as shown in (b), devices may lock to their AES-format audio inputs, although they may only do so when recording. For synchronous replay, devices normally require a separate external sync input, unless they have been modified to lock to the digital audio input on replay as well.

Alternatively, in a small studio, it may be uneconomical or impractical to use a separate SPG. In such cases one device in the studio must be designated as the master. This device would then effectively act as the SPG, operating in the internal sync mode, with all other devices operating in the external sync mode and slaving to it (see Figure 6.3). If a digital mixer were to be used then it could be used as the SPG, but alternatively it would be possible to use a tape recorder, disk system or other device with a stable clock. In such a configuration it would be necessary either to use AES/EBU interfaces for all interconnection (in which case it would be possible to operate in the 'genlock' mode described above, where all devices derive a clock from their digital audio inputs) or to distribute a separate word clock or AES sync signal from the master. In the genlock configuration, the danger exists of timing errors being compounded as delays introduced by one device are passed serially down the signal chain.

In situations where sources are widely spread apart, perhaps even being fed in from remote sites (as might be the case in broadcast operations) the distribution of a single sync reference to all devices becomes very difficult or impossible. In such cases it may be necessary to retime external 'wild' feeds to the house sync before they can be connected to in-house equipment and for this purpose a sampling frequency synchronizer should be used (see section 6.5).

6.4 Clock accuracy considerations

As mentioned above, there are a number of different considerations concerning clock accuracy and the importance of stability that apply at different points in the signal chain. There is the accuracy of the audio sample clock used for A/D and D/A conversion, which will have a direct effect on sound quality, and there is the accuracy of the external reference signal. There is also the question of timing stability in digital audio signals that have travelled over interconnects such as those described in this book, which may have suffered distortions of various kinds. Because the audio sample clock used for conversion in externally synchronized systems must be locked in some way either to the digital audio input clock or the sync reference, it is common for instabilities in either of these signals to affect the stability of the audio sample clock (although they need not). This depends on how the clock is extracted from the digital input signal and the nature of the timing error. Furthermore, clock instability resulting from distortion and interference in the digital interface makes the signal more difficult to decode.

6.4.1   Causes and effects of jitter on the interface signal

Timing irregularities may arise in signals transferred over a digital audio interface due to a number of factors. These may include bandwidth limitations of the interconnect and the effects of induced noise and other signals. Furthermore, the transmitted signal may already have some jitter, either because the source did not properly reject incoming jitter from an input signal, or because its own free-running clock was unstable. AES3 originally specified that data transitions on the interface should occur within ±20 ns of an ideal jitter-free clock. In real products it is normal for interface transition jitter to be well below 20 ns, but when devices which pass on input jitter to their digital outputs are cascaded it is possible for specifications to exceed this value after a number of stages. The degree to which a device passes on jitter is known as its jitter transfer function.

AES3-1997 is somewhat more specific in relation to jitter, specifying limits for output jitter, intrinsic jitter, jitter gain and input jitter tolerance. Output jitter is the sum of the intrinsic jitter of the device's output and that passed through from any timing reference. The intrinsic jitter is specified to be no greater than 0.025 UI (unit interval) when measured using a standard high-pass measurement filter specified in the standard. A UI is specified as the smallest timing unit in the interface and there are 128 UIs per frame (the modulation method can involve transitions in the middle of a bit cell). Sinusoidal jitter gain (the ratio of jitter amplitude at the output of the device to that present at the sync signal input) should be no greater than 2 dB, again measured using a standard filter. Input jitter tolerance (the maximum jitter value below which a device should correctly decode input data) is specified as 0.25 UI peak-to-peak above 8 kHz rising to 10 UI below 200 Hz.

The received signal from a standard two-channel AES interface will have an eye pattern that depends on amplitude and timing irregularities (see Chapter 4). Amplitude errors will close the eye vertically and timing errors will close it horizontally. The limits for correct decoding are laid out in the specification of the interface. Nonetheless, some receivers are better than others at decoding data with a poor eye pattern and this has partly to do with the frequency response of the phase-locked loop in the receiver and its lock-in range (see below). It also depends on the part of the signal from which the decoder extracts its clock, since some transitions are decidedly more unstable than others when the link is poor. Decoders that are very tolerant of poor input signals may at the same time be bad at rejecting jitter. Although they may decode the signal, the resulting sound quality may be poor if the signal is converted within the device without further rejection of jitter and the device may pass on excessive jitter at its output.

Dunn[4] and Dunn and Hawksford[6] have both carried out simulations of the effects of link bandwidth reduction on standard two-channel interface signals. The important conclusions of their work are as follows. When a link suffers high-frequency loss there will be a reduction in amplitude of the shorter pulses and a slowing in rise and fall times at transitions, the effect of which is to delay the zero-crossing transition after short pulses less than the delay after longer pulses. This variable delay is effectively jitter and is solely a result of high-frequency loss. Figure 6.4 shows the HF loss model which was used in both studies (which, although simplistic, is considered a good starting point for analysis), the time constant of which is *RC*. Figure 6.5 shows a comparison between simulated bi-phase mark data at time constants of 200 ns and 50 ns, showing clearly that at 200 ns the shorter pulses are more attenuated than the longer (a time constant of 200 ns corresponds to a roll-off of -3 dB at 0.8 MHz, whereas 50 ns corresponds to a similar roll-off at around 3.18 MHz). Dunn's results show clearly that for links with a bandwidth of less than around 3 MHz the jitter suffered by transitions in the main part of the AES subframe (the audio data time slots) is far greater than that suffered by the penultimate transition of the Y preamble, as shown in Figure 6.6. Dunn and Hawksford also provide convincing evidence that the jitter is highly correlated with the audio signal and is affected by the difference between the number of zeros and ones in the signal. This situation clearly becomes more critical at higher data rates than those originally specified.

6.4.2 Audio sampling frequency

It is important to separate the discussion of long-term sample frequency accuracy from stability in the short term. Short-term instability is called 'jitter' and long-term inaccuracy would manifest itself as drift (if in one direction only), or wow and flutter (if cyclically modulated) in extreme cases. Jitter will be covered in the next section.

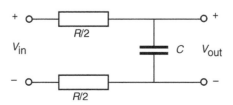

Figure 6.4 High-frequency loss model used in simulating the effects of cables.

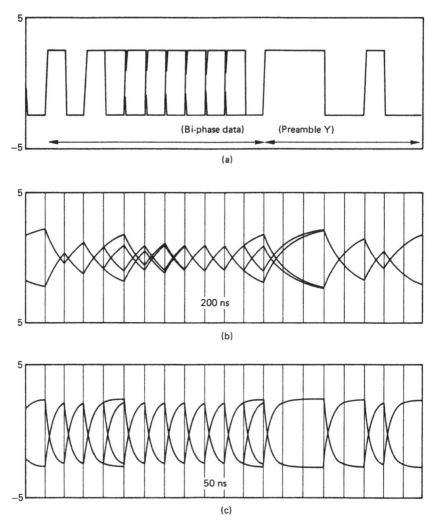

Figure 6.5 A simulation of the effects of high-frequency loss on an AES3-format signal. At (a) the original signal is shown (both polarities are superimposed for clarity), whilst (b) and (c) show the data waveform at link time constants of 200 ns and 50 ns respectively. (Reproduced from J. Dunn, with permission.)

For professional equipment, the nominal sampling frequency should be accurate to within ±10 ppm if it conforms to AES5 recommendations (although the standard only explicitly states this at 48 kHz). This corresponds to an allowable peak drift in the sample period of ±0.21 ns at a sampling frequency of 48 kHz, but implies nothing about the *rate* at which that modulation takes place. When a device 'free runs' it is locked to its own internal oscillator, which for fixed sampling frequencies is normally

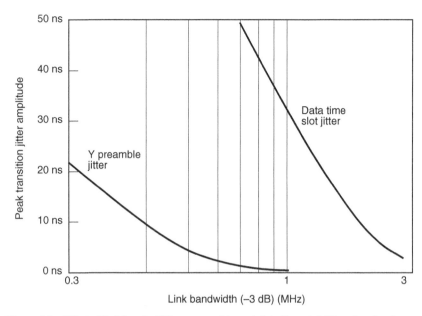

Figure 6.6 Effect of link bandwidth on preamble and data time slot jitter, showing how the Y preamble is affected much less by reduced bandwidth than the data edges. (Reproduced from J. Dunn, with permission.)

a crystal oscillator capable of high accuracy. However, in variable-speed modes a crystal oscillator cannot easily be used and some form of voltage-controlled or other oscillator may take its place, having a less stable frequency. In such cases a device is not expected to meet the stability requirements of AES5 or AES11.

In consumer equipment the sampling frequency is normally less carefully specified and controlled, often making it difficult to interconnect consumer and professional equipment without the use of a sampling frequency convertor or a synchronizer. IEC 60958 specifies three levels of sampling frequency accuracy: Level I ('high' accuracy) = ±50 ppm; Level II (normal accuracy) = ±1000 ppm; and Level III (variable pitch shifted clock mode), which is undefined except to say that it can only be received by special equipment and that the frequency range is likely to be ±12.5% of the nominal sampling frequency. Again nothing is said about the rate of sample clock modulation. Consumer sampling frequency and clock accuracy are indicated in bits 24–27 and 28–29 of channel status in the digital audio interface signal (see section 4.8). There is no such indication in professional channel status, except in AES11-type reference signals (see above).

At the professional limit of ±10 ppm, a nominal sample clock of 48 kHz could range over the limits 47 999.52 Hz to 48 000.48 Hz – a speed tolerance of 0.001% – whereas a normal accuracy consumer device at the same nominal sampling frequency could be anything from 47 956 Hz to 48 048 Hz – a speed tolerance of ±0.1%.

6.4.3 Sample clock jitter and effects on sound quality

Short-term timing irregularities in sample clocks may affect sound quality in devices such as A/D and D/A convertors and sampling frequency convertors. This is due to modulation in the time domain of the sample instant (see section 2.7.4), resulting in low-level signal products within the audio spectrum. The important features of jitter are its peak amplitude and its rate, since the effect on sound quality is dependent on both of these factors taken together. Shelton[3], by calculating the rms signal-to-noise ratio resulting from random jitter, showed that timing irregularities as low as 5 ns may be significant for 16-bit digital audio systems over a range of signal frequencies and that the criteria are even more stringent at higher resolutions and at high frequencies. The effects are summarized in Figure 6.7.

When jitter is periodic rather than random, it results in the equivalent of 'flutter', and the effect when applied to the sample clock in the conversion of a sinusoidal audio signal is to produce sidebands on either side of the original audio signal due to phase modulation, whose spacing is equal to the jitter frequency. Julian Dunn[4] has shown that the level of the jitter sideband (R_j) with relation to the signal is given by:

$$R_j \text{ (dB)} = 20\log (J\omega_i/4)$$

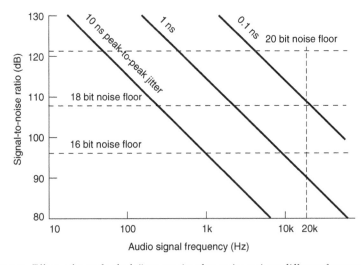

Figure 6.7 Effects of sample clock jitter on signal-to-noise ratio at different frequencies, compared with theoretical noise floors of systems with different resolutions. (After W. T. Shelton, with permission.)

where J is the peak-to-peak amplitude of the jitter and ω_i is the audio signal frequency. Using this formula he shows that for sinusoidal jitter with an amplitude of 500 ps, a maximum level 20 kHz audio signal will produce sidebands at -96.1 dB relative to the amplitude of the tone.

What is important, though, is the *audibility* of jitter-induced products and Dunn[4,5] attempted to calculate this based on an analysis of the resulting spectrum using accepted audibility curves, based on critical band masking theory, assuming that the audio signal is replayed at a high listening level (120 dB SPL). As shown in Figure 6.8, which plots jitter amplitude against jitter frequency (not audio frequency) for just-audible modulation noise on a worst-case audio signal, the jitter amplitude may in fact be very high (>1 µs) at low jitter frequencies (up to around 250 Hz) because the sidebands will be masked at all audio frequencies, but the amount allowed falls sharply above this jitter frequency although it may still be up to ±10 ns at jitter frequencies up to 400 Hz.

The original version of AES11 specified tolerances for jitter on the sampling frequency clock, but this was dropped in the 1997 revision in favour of a statement to the effect that the clock tolerance requirements for A/D and D/A conversion would have to be more stringent than that for a Grade 1 reference signal in respect of random jitter and jitter modulation. This reinforces the point that sampling clock jitter and interface clock jitter are related but different problems. The effect of jitter on sampling

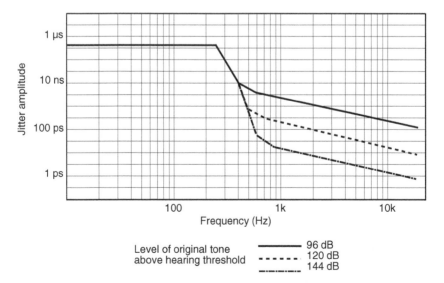

Figure 6.8 Sample clock jitter amplitude at different frequencies required for just-audible modulation noise on a worst-case audio signal. (After J. Dunn, with permission.)

clocks and the audible result thereof has become a large and complex subject and it is not proposed to deal with it further in this book, the primary purpose of which is to cover interfacing issues. A comprehensive study by Chris Dunn and Malcolm Hawksford[6] attempted to survey the effects of interface-induced jitter on different types of DAC and this paper warrants close study by those whose business it is to design high quality DACs with digital audio interfaces.

The implication of section 6.4.1 is that it is greatly preferable to derive a stable sample clock from one of the reliable preamble transitions than from the audio data slot transitions, although there is evidence that many devices do use the data transitions. One older interface receiver chip adjusted its PLL on every negative-going transition in the interface signal, for example. Since transitions in the audio data part of the subframe are not only more sensitive to line-induced jitter, but also determined by the audio signal, it is even possible for a signal on the B audio channel to modulate the sampling clock such that jitter sidebands appear in the A channel that are tonally related to the signal in the B channel.

The rejection of jitter by the receiver depends principally on the frequency response of its PLL and in general the narrower the response of the PLL and the lower its cut-off frequency the better the rejection of jitter. The problem with this is that such PLLs may not lock up as quickly as wide bandwidth PLLs and may not lock over a particularly wide frequency range. A solution suggested by Dunn and Hawksford is to use a wideband PLL in series with a low bandwidth version which switches in after conditions have stabilized. Alternatively RAM buffering may be used between the interface decoder and the convertor, the data being clocked out of the buffer to the convertor under control of a more stable clock source. In a hi-fi system it is possible that the stable clock source's frequency could be independent of the incoming data clock, provided that the buffer was of sufficient size to accommodate the maximum timing error between the two over the duration of, say, a CD, but the better solution adopted by some manufacturers in two-box CD players is to generate a synchronizing clock in the convertor which is fed back to reference the speed of the CD transport and thus the rate of data coming over the digital interface (see Figure 6.9).

In a professional digital audio system where all devices in the system are to be locked to a common sampling frequency clock it would normally be necessary to lock any convertors to an external reference. An AES11-style reference signal derived from a central generator might have suffered similar degradations to an audio signal travelling between two devices and thus exhibit transition timing jitter. In such cases, especially in areas where high quality D/A conversion is required, it is advisable either

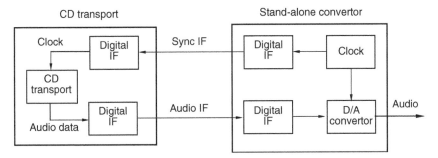

Figure 6.9 In one high quality CD player with a separate convertor, rather than deriving the convertor clock from the incoming audio data, the convertor clock is fed back to the transport via a separate interface in order to maintain synchronization of the incoming data.

to reclock the reference signal or to use a local high quality reference generator, slaved to the central AES11 generator, with which to clock the convertors.

6.5 Use and function of sampling frequency synchronizers

A sampling frequency synchronizer may be used to lock a digital audio signal to a reference signal. It could also perform the useful function of reclocking distorted remote signals to remove short-term timing errors (jitter). Three main approaches to sampling frequency synchronization are necessary, depending on the operational requirement. Each will be discussed in turn. These are:

(a) *Frame alignment*: to deal with signals which are of identical sampling frequency but which are more than ±25% of a frame period out of phase with the reference.
(b) *Buffering*: to deal with signals of nominally the same sampling frequency but which are not locked to the same reference and thus drift slightly with relation to each other.
(c) *Sampling frequency conversion*: to deal with signals whose sampling frequencies differ by a larger amount than implied in (b), such as between 44.1 and 48 kHz, or between consumer and professional systems in which the consumer device's sampling rate is nominally the same as the professional device's but within a large tolerance (see section 6.4 above).

6.5.1 Frame alignment

It may be necessary to correct for timing errors in signals that are synchronous to the master clock but have travelled long distances. These will have been delayed and so be out of phase with the reference. This is more properly referred to as 'frame alignment' and is only necessary when a signal is more than ±25% of a frame period delayed with reference to the sync reference. Propagation delays are not great, however: for example, an AES/EBU signal must travel some 3.7 km down a typical cable before it is delayed by one sample period; thus it is most unlikely that such a situation will arise in real operational environments unless a large static phase error has been introduced in the sample clock of an incoming signal due to it having been cascaded through a number of devices operating in the genlock mode (see section 6.3).

In order to conform to AES recommendations, frame alignment should rephase the signal to bring it within ±5% of a frame period compared with the sync reference. Input signals less than ±25% adrift are also expected to be brought within this ±5% limit at the output, but this is normally performed within the device itself. Reframing of signals more than ±25% adrift may be performed within the device, but if not an external reframer would be required.

6.5.2 Buffering

For signals of nominally the same frequency but very slightly adrift it is possible to use a simple buffer store synchronizer, such as those described by Gilchrist[7] and also by Parker[8]. In this type of synchronizer, a typical block diagram of which is pictured in Figure 6.10, audio samples are written sequentially into successive addresses of a solid state memory configured in the FIFO manner. These samples are read out of the memory a short time later, clocked by the reference signal, the buffer providing a short-term store to accommodate the variation in input and output rates. If the output rate is slightly faster than the input rate then the buffer will gradually become empty and if it is slower than the input rate the buffer will gradually become full, requiring action at some point to avoid losing data or repeating samples because the buffer cannot be infinitely large. At such a time the read address is reset to the mid point of the buffer, resulting in a slight discontinuity in the audio signal. This discontinuity may be arranged to occur within silent passages of the programme, or alternatively a short crossfade may be introduced at the reset point to 'hide' the discontinuity.

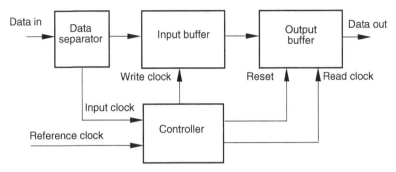

Figure 6.10 Block diagram of an example of a simple buffer store synchronizer.

Buffer store synchronizers have the advantage that most of the time the audio signal is copied bit for bit between input and output, with discontinuities only occurring once every so many minutes, depending on the size of the buffer and the discrepancy in input and output sampling rates. The larger the buffer the longer the gaps between buffer resets, or the greater the discrepancy between sampling rates which may be accommodated. The price for using a larger buffer is a longer delay between input and output, and this must be chosen with the operational requirement in mind. Using a buffer store capable of holding 480 samples, for example, a delay of around 5 ms would result, and buffer resets would occur every 8.3 minutes if the sampling frequencys were at the extremes of the AES5 tolerance of ± 10 ppm.

6.5.3 Sampling frequency conversion

For signals whose sampling frequencies differ by too great an amount to be handled by a buffer store synchronizer it will be necessary to employ sampling frequency conversion. This can be used to convert digital interface signals from one rate to another (say from 44.1 to 48 kHz) without passing through the analog domain. Sampling frequency conversion is not truly a transparent process but modern convertors introduce minimal side effects.

The most basic form of sampling frequency conversion involves the translation of samples at one fixed rate to a new fixed rate, related by a simple fractional ratio. Fractional-ratio conversion involves the mathematical interpolation of samples at the new rate based on the values of samples at the old rate. Digital filtering is used to calculate the amplitudes of the new samples such that they are mathematically correct based on the impulse response of original samples, after low-pass filtering with an upper limit of the Nyquist frequency of the original sampling rate. A clock rate common to both sampling frequencies is used to control the interpolation process. Using this method, some output samples will coincide with input samples, but only a limited number of possibilities exist for the interval between input and output samples. Such a process is nominally jitter free.

If the input and output sampling rates have a variable or non-simple relationship, output samples may be required at any interval in between input samples. This requires an interpolator with many more clock phases than for fractional-ratio conversion, the intention being to pick a clock phase which most closely corresponds with the desired output sample instant at which to calculate the necessary coefficient. There will clearly be a timing error, the audible result of which is equivalent to the effect of jitter, and this may be made smaller by increasing the number of possible interpolator phases. If the input sampling rate is continuously varied (as it might be in variable-speed searching or cueing) the position of interpolated samples with relation to original samples must vary also, and this requires real-time calculation of filter phase.

Errors in sampling frequency conversion should be designed so as to result in noise modulation below the noise floor of a 16-bit system and preferably lower. For example, one such convertor is quoted as introducing distortion and noise at $-105\,dB$ ref. 1 kHz at 0 dB FS (equivalent to 18-bit noise performance).

6.6 Considerations in video environments

Audio and video/film have traditionally required synchronization for the purposes of achieving lip sync. Film and video are both discrete in that they have frames, and when audio was analog synchronizing to sufficient accuracy for lip sync caused no difficulty. Now that audio is also digital, it too is made up of discrete information and more accurate synchronizing with video becomes necessary. In environments where digital audio is used with video signals it is important for the audio sampling rate to be locked to the same master clock as the video reference. The same applies to timecode signals which may be used with digital audio and video equipment. A number of proposals exist for incorporating timecode within the standard two-channel interface, each of which has different merits, and these will be discussed below.

6.6.1 Relationships between video frame rates and audio sampling rates

People using the PAL or SECAM television systems are fortunate in that there is a simple integer relationship between the sampling frequency of 48 kHz used in digital audio systems for TV and the video frame rate of 25 Hz (there are 1920 samples per frame). There is also a simple relationship between the other standard sampling frequencies of 44.1 and 32 kHz and the PAL/SECAM frame rate, as was shown in Table 4.2. Users of NTSC TV systems (such as the USA and Japan) are less fortunate because the TV frame rate is 30/1.001 (roughly 29.97) frames per second, resulting in a non-integer relationship with standard audio sampling frequencies. The sampling frequency of 44.056 kHz was introduced in early digital audio recording systems that used NTSC VTRs, as this resulted in an integer relationship with the frame rate. For a variety of historical reasons it is still quite common to encounter so-called 'pull-down' sampling frequencies in video environments using the NTSC frame rate, these being 1/1.001 times the standard sampling frequencies, which mainly serves to complicate matters.

The standard two-channel interface's channel status block structure repeats at 192 sample intervals, and in 48 kHz systems there are exactly ten audio interface frames per PAL/SECAM video frame, simplifying the synchronization of information contained in channel status with junctures in the TV signal and making possible the carrying of EBU timecode signals in channel status as described below.

As described by Shelton[9] and others[10,11] it is desirable to source a master audio reference signal centrally within a studio operation, just as a video reference is centrally sourced. These two references are normally locked to the same highly stable rubidium reference, which in turn may be locked to a standard reference frequency broadcast by a national transmitter. As shown in Figure 6.11, the overall master clock contained in a central apparatus room will be used to lock the video SPG (distributed as 'black and burst' video), the audio reference signal generator (distributed as a standard AES11 reference signal) and the colour subcarrier synthesizer. The video reference is used in turn to lock timecode generators. Appropriate master clock frequencies and division ratios may be devised for the application in question.

The sync point between audio and video reference signals defined in AES11 is the half amplitude point of the leading edge of line sync of line 1 of the video frame in systems where there is an integer number of AES frames per video frame. This is synchronized to the start of the X or Z

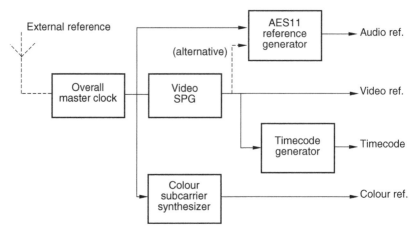

Figure 6.11 In video environments all audio, video and timecode sync references should be locked to a common clock.

audio preamble. The situation is complicated with NTSC video since there is not an integer number of audio samples per frame. The desired alignment of audio preamble and video line sync only occurs once every five frames and an indicator is supposed to be included in the video reference to show which frame acts as the sync point.

An alternative approach to video sync is found on some audio equipment, especially digital tape recorders. Here the audio device accepts a video sync reference and derives its sample clock by appropriate multiplication and division of this reference internally. DIP switches may be provided on the audio device to select the appropriate frame rate so that the correct ratio results.

6.6.2 *Referencing of VTRs with digital audio tracks*

Video tape recorders (VTRs), both analog and digital, are often equipped with digital audio tracks. Digital VTRs are really data recorders with audio and video interfaces. The great majority of the data is video samples and the head rotation is locked to the video field timing. Part of each track is reserved for audio data, which uses the same heads and much of the same circuitry as the video on a time-division basis. As a result the audio sampling rate has to be locked to video timing so that the correct number of audio and video samples can be assembled in order to record a track.

At the moment the only audio sampling rate supported by digital VTRs is 48 kHz. This causes no difficulty in 625/50 systems as there are exactly 960 sample periods in a field and a phase-locked loop can easily multiply the vertical rate to produce a video synchronous 48 kHz clock. Alternatively, line rate can be multiplied by 384/125. However, the 0.1% offset of 525 line systems makes the actual field rate 59.94 Hz. The fields are slightly longer than at 60 Hz and in 60 fields there will be exactly 48 048 sample periods. Unfortunately this number does not divide by 60 without a remainder. The smallest number of fields which contain a whole number of samples is five. This makes the generation of the audio sampling clock more difficult, but it can be obtained by multiplying line rate by 1144/375.

When a DVTR is recording using analog audio inputs, the heads rotate locked to input video, which also determines the audio sampling rate for the ADCs in the machine and there is no difficulty. On replay, the synchronism between video and audio is locked in the recording and the audio sampling rate on the output will be locked to station reference. Any receiving device will need to slave to the replay audio sampling rate. On recording with a digital audio input, it is necessary that the digital audio source is slaved to the input video of the DVTR. This can be achieved either by taking a video-derived audio sampling rate reference from the DVTR to the source, or by using a source which can derive its own sampling rate from a video input. The same video timing is then routed to both the source and the DVTR.

If these steps are not followed, there could be an audio sampling rate mismatch between the source and destination and inevitably periodic corruption will occur. With modern crystal-controlled devices it is surprising how long an unlocked system can run between corruptions. It is easy mistakenly to think a system is locked simply because it works for a short whilst when in fact it is not and will shows signs of distress if monitored for longer.

In some VTRs with digital audio tracks it has been possible for there to arise a phase relationship between the digital audio outputs of a VTR and the video output which is different each time the VTR is turned on, causing difficulties in the digital transfer of audio and video data to another VTR. When the phase relationship is such that the incoming digital audio signal to a VTR lies right on the boundary between two sample slots of its own reference there is often the possibility of sample slips or repeats when timing jitter causes vacillation of sync between two sample periods. This highlights the importance of fixing the audio/video phase relationship.

Manufacturers of video recorders, though, claim that the solution is not as simple as it seems, since in editing systems the VTR may switch between different video sync references depending on the operational mode. Should the audio phase always follow the video reference selection of the VTR? In this case audio reframing would be required at regular points in the signal chain. A likely interim solution will be that at least the phase of the house audio reference signal with relation to the house video sync will be specified, providing a known reference sync point for any reframing which might be required in combined audio/video systems.

6.6.3 Timecode in the standard two-channel interface

In section 4.8 the possibility of including 'sample address' codes in the channel status data of the interface was introduced. Such a code is capable of representing a count of the number of samples elapsed since midnight in a binary form, and when there is a simple relationship between the audio sampling frequency and the video frame rate as there is with PAL TV signals, it is relatively straightforward to convert sample address codes into the equivalent value in hours, minutes, seconds and frames used in SMPTE/EBU timecode.

At a sampling frequency of 48 kHz the sample address is updated once every 4 ms, which is ten times per video frame. At NTSC frame rates the transcoding is less easy, especially since NTSC video requires 'drop-frame' SMPTE timecode to accommodate the non-integer number of frames per second. A further potential difficulty is that the rate of update and method of transcoding for sample address codes is dependent upon the audio sampling frequency, but since this is normally fixed within one studio centre the issue is not particularly important.

Discussions ran for a number of years on an appropriate method for encoding SMPTE/EBU timecodes in a suitable form within the channel status or user bits of the audio interface. At the time of writing there is no published standard for this purpose, although a number of European broadcasters have adopted the non-standard procedure described below. There is also potential for incorporating timecode messages into the HDLC packet scheme determined for the user bit channel in AES18 (see section 4.7.1). The advantage of this option is that the data transfer rate in AES18 is independent of the audio sampling frequency, within limits, so the timecode could possibly be asynchronous with the audio, video, or both, if necessary. Such a scheme makes it difficult to derive a phase reference between the video frame edge and the timecode frame edge, but additional bits are proposed in this option to indicate the phase offset between the two at regular points in the video frame.

An important proposal was made and adopted by a number of European broadcasters[12], which replaced the local sample address code in bytes 14–17 of channel status with four bytes of conventional SMPTE/EBU timecode. These bytes contained the BCD (Binary Coded Decimal) time values for hours, minutes, seconds and frames, as replayed from the device in question, in the following manner:

Byte 14 = frames (a four-bit BCD value for both tens and units of frames)
Byte 15 = seconds (ditto for seconds)
Byte 16 = minutes (ditto for minutes)
Byte 17 = hours (ditto for hours)

The time-of-day sample address bytes (18–21) are replaced by time-of-day timecode in the same way. At 48 kHz with EBU timecode the time-code value is thus repeated ten times per frame.

6.7 Compatibility issues in audio interfacing

Despite the many standards relating to audio interfaces, or perhaps because of them, it is possible that practical difficulties may arise when attempting to interconnect two or more devices. The problem of 'getting devices to talk to each other' is possibly less serious than it was when this book was first written, but there are still areas of difficulty and the number of interface types is large. Communication problems may usually be boiled down to one of a few common sources of incompatibility. Not only must the user know how to work around basic incompatibilities, but also it is necessary to be aware of the possibility for incorrect communication. Devices may be made to 'talk' but something may be lost or gained in the translation!

The majority of this section is devoted to the standard two-channel interface. Communications between devices using dissimilar manufacturer-specific interfaces will nearly always require the use of a format convertor such as those described in section 6.12.1. In addition to the extended discussion of practical interfacing contained in the main text, there is a reference 'troubleshooting guide' to audio interfacing at the end of this section. If problems are encountered with communication over computer networks one normally needs to deal with them as computer networking problems rather than audio interfacing problems. This may involve correct configuration of routers, IP addresses, network drivers and ports, among other things. This is a topic in its own right, about which much is written, and will not be covered further here.

6.7.1 Incompatibilities between devices using the standard two-channel interface

There are only really two reasons why devices using nominally the same interface will not communicate, summed up as either electrical incompatibility or data incompatibility. There is also, of course, the possibility for sampling frequency incompatibility between devices, but this could be considered as a combination of electrical and data incompatibility. If the two devices are using the same electrical interface, such as AES3 for professional purposes or IEC 60958–3 for consumer purposes, there is only a very small chance of electrical mismatch, particularly now that the more recent versions of the standards have limited the options for electrical incompatibility. The more likely cause of any problems is that differences exist between the data transmitted and that expected by the receiver. If direct links are attempted between consumer and professional equipment then there is potential for both electrical and data incompatibility, as discussed below.

6.7.2 Electrical mismatch in professional systems

Between identical interfaces (transmitter and receiver) the most likely electrical problems to arise are (a) loss over long lines; (b) noise and distortion over long lines; (c) impedance mismatch. These can in general be avoided by good system design and by adhering to the recommendations contained in the appropriate standard, but occasionally one encounters a poor electrical installation or needs to make use of existing wiring which may not be ideal for the job of carrying digital audio. When such electrical problems are encountered the most likely symptom is that the receiver will find it difficult or impossible to lock to the incoming data signal, resulting in intermittent operation or indication of 'loss of lock' at the digital input. In practice receivers vary widely in their ability to lock to poor quality signals, and thus it may be found that a signal which works satisfactorily with one receiver proves unsatisfactory with another. Using devices such as those discussed in section 6.9 it is possible to determine the 'health' of the received data signal, as well as examining problems with data.

A receiver conforming to AES3 should be able to decode a signal with a minimum eye height of 200 mV, and since the transmitter normally produces at least 5 volts the resistive cable attenuation has to be quite large before it will reduce the eye height below this value. More problematical than simple resistive loss is high-frequency roll-off over a long cable. This may need to be corrected by using suitable equalization at the receiver (see section 4.3.3), although equalization should be treated with care since it will only work if the line is relatively noise free. (If the line is noisy then equalization may actually make the problem worse and it has been suggested that if such equalization is to be used it perhaps belongs at the transmitter end rather than the receiver.)

High-frequency loss affects the narrow pulses of the data stream before the wide ones and these narrow pulses may either fail to provide a zero crossing or even disappear altogether in extreme cases. The greater the HF loss the more likelihood there is of intersymbol interference and data edge timing jitter, making it more difficult for the receiver to lock to the received data. Dunn[1] suggests that the cable used should not exhibit attenuation at 6 MHz which is more than 6 dB greater than that at 1 MHz over the distance used, otherwise equalization will be required (this is for basic sampling rates). Operating the interface at higher data rates than the original specification will put greater demands on this criterion, extending it to higher frequencies. Although conventional analog audio cable can and has been used in many cases, it is becoming common for new installations to be wired for digital audio with cable having higher specifications and a characteristic impedance which is better controlled and closer to 110 ohms than conventional microphone cable.

Impedance mismatches were more likely under the original AES3 specification than they became under the 1992 revision, due to the 250 ohm termination impedance specified in AES3-1985. (This was to allow for between one and four receivers to be connected in parallel across one signal line.) Such mismatches may result in internal reflections such that transmitted pulses are reflected from the receiving end to interfere with pulses travelling in the opposite direction, and the cable may begin to function as an antenna – both picking up and radiating interference. Now that the termination impedance is specified at 110 ohms and point-to-point interconnects are required, it may be necessary to modify the input impedance of older receivers by fitting a parallel resistor of around 200 ohms so as to make the termination nearer to the correct value. If it is necessary to feed more than one receiver from a single driver it is recommended that suitable digital distribution amplifiers (DDAs) are used, or alternatively one could use passive resistive splitters.

In order to avoid impedance mismatches in between transmitter and receiver it is important that the cable used is consistent along its length and that joins are not made between dissimilar cable types. Problems may also arise if digital audio signals are routed via analog patch bays in which short sections of cabling with mismatched impedances may exist. Any cable 'stubs' in such installations produce short-delayed reflections with fairly high amplitude that can interfere with the transmitted data signal.

Users wishing to carry signals over long distances may wish to consider the possibility of adopting the 75 ohm unbalanced interconnect specified in AES3-ID or SMPTE 276M as an alternative to balanced 110 ohm interfacing. Convertors are available which perform the job very easily. It has been suggested by a number of sources that HF interference such as RF sources is better rejected by the effectiveness of cable screening than by the balance of the electrical interface, and that 75 ohm coax cable has better controlled impedance than audio microphone cable.

A final point to bear in mind is that although transformers are not mandatory in most versions of the two-channel interface standard they may be used to ensure good electrical isolation and earth separation where appropriate. The transformer is standard in the EBU version of the interface, since it was regarded as important in broadcast studio centres where earth continuity is normally avoided between operational areas.

6.7.3   Data mismatch in professional systems

Data mismatch between professional devices using the standard two-channel interface has typically been confined to problems with the implementation of channel status (see section 4.8), but there is also the possibility for differences in sampling rate and audio word length, as discussed below. In more recent systems there are numerous options for higher sampling frequencies, operating the interface in single-channel-double-sampling frequency mode or simply increasing the data rate, which can lead to difficulties in communication. There is also the increased likelihood that interfaces may be operated in the 'non-audio' or 'other purposes' mode for carrying data-reduced audio signals, as described in Chapter 4. User bit incompatibilities might arise if greater use was made of this channel, but few current professional devices take any notice of the state of the user bit. The validity bit is historically another root of trouble and its handling was discussed in section 4.6. As discussed in section 6.12, a number of manufacturers now produce devices specifically designed to analyse and/or correct for data incompatibilities between pieces of equipment and such 'fix-it' boxes can be very useful in encouraging communications between systems.

Incompatibilities in channel status implementation can give rise to a variety of symptoms ranging from complete failure of communication to seemingly correct but actually improper communication. The reason that such incompatibilities have arisen is largely that the original AES3 specification was less than specific on how devices should set channel status bits that were not used, as well as how they should respond when receiving data that they were incapable of handling. Consequently all sorts of channel status implementations exist in commercial products, although it should be said that most of the time devices communicate without problems. Because of these potential difficulties, AES3-1992 was more specific about channel status implementation, specifying three levels of implementation depending upon the application (see section 4.8.4). As a result there should be less difficulty between modern devices than between older ones.

The most common problem areas in channel status implementation were historically (a) in the CRCC byte (byte 23); (b) in the signalling of pre-emphasis; (c) in the setting of the consumer/professional flag; and (d) in the indication of sampling rate. Less common problems arose when the left and right channels had different channel status data (making it difficult to decide which was correct) or where the channel status block was the wrong length (one famous example exists of a 191-byte channel status block!).

The problem with the CRCC byte was that not all transmitting devices included it at the end of the channel status block and some early devices implemented it incorrectly, thereby confusing those that expected to see the correct CRCC data. Receivers that check the CRCC will indicate an almost continuous CRC error when decoding an input signal that does not contain CRCC or where it is incorrectly implemented. The reaction of such a receiver will vary from complete refusal to accept the signal to acceptance of the signal whilst flagging a CRC error. It is impossible to state what the 'correct' response should be in such a case, as there is no way of telling whether the channel status data is correct or not. Interface signal processors such as the 'fix-it' boxes mentioned above often perform the useful function of inserting correct CRCC data into the channel status blocks of signals that lack it. It might reasonably be assumed that a device seeing CRCC bytes repeatedly set to zero would assume that it was not in use and thereafter ignore it, but this is rarely the case at present.

The type of pre-emphasis used should be indicated in bits 2–4 of byte 0 of channel status and it is possible that emphasis may have been applied to a signal without indicating this in channel status. Such incorrectness will not prevent the interface from working but may give rise to a pre-emphasized signal being carried through further stages in the signal chain without being de-emphasized. The only way to tell if a signal is pre-emphasized (when it is not indicated) is really to listen to it, since pre-emphasized signals will have an exaggerated HF response. The correct pre-emphasis flags may be set using a suitable interface processor, or the signal may be de-emphasized in the digital domain and the flags set to the no emphasis state.

Apart from the consumer/professional flag (discussed below), the sampling frequency indication is the other main area of difficulty in channel status. Not all devices indicate the sampling frequency of the signal in bits 6 and 7 of byte 0, and this can cause a lack of communication when received by a device expecting to see such data. Japanese devices in particular often will not accept data if it does not have the sampling frequency flags set correctly. There is also now the additional indication of sampling frequency in byte 4 of channel status (AES3, Amendment 3 – 1999) to complicate matters, although this is not a requirement for correct functioning of the interface. It is difficult for a receiver to know what to do when presented with a sampling rate flag that contradicts the true audio sampling rate. In such cases it might be suggested that the device should rely on its detection of the true rate rather than the indicated rate, whilst perhaps flagging a mismatch on the front panel. Some confusion also exists over the interpretation of bit 5, byte 0, which indicates 'source sampling frequency unlocked'. The question is 'with reference to what is the sampling frequency unlocked?' – the internal clock, an external reference? When this bit is set to '0' (the default state) nothing can be concluded about the locked state of the sample clock and many systems do not set this bit anyway. When set to '1' all one can say is that there is some problem with the lock of the sample clock and that its frequency may not be relied upon. In systems synchronized to a master reference signal its presence could be used to indicate that there was a free-running clock in a source device earlier in the signal chain.

6.7.4   Electrical mismatch between consumer and professional systems

As described in Chapter 4 there is so much similarity between the consumer and professional interfaces that it is tempting to think that consumer devices can be connected directly to professional systems or vice versa. Indeed there is a strong operational motive for this because consumer and professional digital equipment are often used together in studios and programme material is often copied between systems. The problem is that although it is possible in some cases to make the electrical interconnection work, there are other difficulties to contend with such as the almost total dissimilarity in channel status and user bits. There is also the possibility that the consumer device may have a much less stable sample clock and be unable to lock to an external reference, giving rise either to problems in decoding or the danger that sample clock jitter will be passed on to other devices in the professional system. Ideally, therefore, one should not attempt the direct interconnection of consumer and professional equipment, preferring rather to use one of the many interface convertors that exist on the market which will set the necessary channel status bits correctly and convert the signal to the new electrical format. It may also be necessary to resynchronize the signal from a consumer device using a buffer store or sample rate convertor (see section 6.5) by locking it to the reference signal of the professional system, thereby ensuring that its sampling rate is the same as that of the professional system and hopefully removing any excessive clock jitter.

For cases in which the only solution is to attempt direct interconnections some guidelines will be given. Clearly such set-ups should be viewed as temporary and allow for the possible incompatibilities in channel status. Consumer-to-professional electrical connection is often possible because the consumer interface peak output voltage is around 0.5 V and the minimum allowed input voltage to a professional system is 0.2 V. Provided that the wire is not too long the signal may be decoded. As shown in Figure 6.12 the centre core of the consumer coaxial lead may be connected to pin 2 of the professional XLR and the shield to pins 1 and 3. Clearly there will be an impedance mismatch and commercial impedance transformers are available to convert between either 75 and 250 ohms or 75 and 110 ohms. One Japanese manufacturer recommends the circuit shown in Figure 6.13 to balance a consumer output for carrying it over longer distances.

Professional-to-consumer connection may also work, again depending on channel status, since a consumer input is not normally damaged by the higher professional signal voltage. Two circuits are shown in Figure 6.14, depending on whether the professional output is transformer balanced (floating) or driven directly from TTL-level chips (balanced, but not floating). It is also possible to convert signals between consumer electrical and optical formats. Suggested circuits are shown in Figure 6.15.

Unfortunately the original IEC 958 did not exclude the possibility of using the Type 2 *electrical* interface with professional *data*, or indeed vice

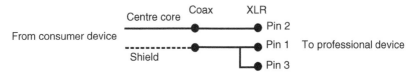

Figure 6.12 Temporary method of interconnection between consumer and professional interfaces.

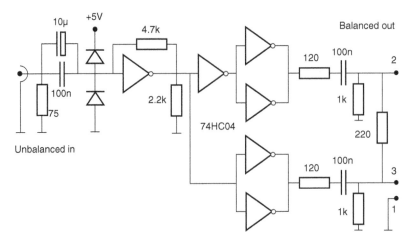

Figure 6.13 An example of a circuit suggested by one manufacturer for deriving a balanced digital output from consumer equipment.

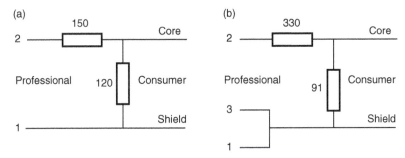

Figure 6.14 Examples of impedance matching adaption circuits between professional and consumer devices: (a) non-floating, and (b) floating (transformer-balanced) professional sources.

versa (the two subjects were entirely separate in the document). This occasionally led to some unexpected implementations. Although one normally expects to find professional data coming out of XLR connectors there are cases where manufacturers or dealers have simply taken a consumer device and provided it with a so-called 'professional' output by feeding the consumer output via an RS-422 driver to an XLR connector. This may be cheap but it is to be discouraged since it leads users to think that the equipment may be connected directly to professional systems, whereas in practice the professional system may refuse to accept it due to the consumer flag being set in the first bit of channel status. If the data is accepted it is possible that further difficulty may arise due to the misinterpretation of channel status data.

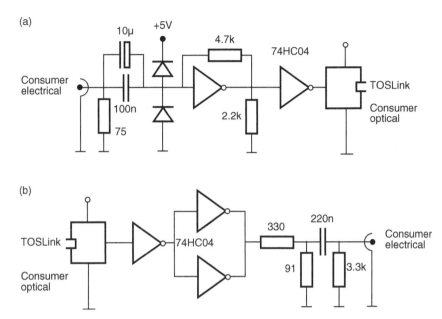

Figure 6.15 An example of a circuit suggested by one manufacturer for converting (a) consumer electrical to consumer optical format, and (b) optical to electrical.

6.7.5 Data mismatch between consumer and professional systems

Although the audio part of the subframe is to all intents and purposes identical between the two interfaces there are key differences in the channel status data that have already been described in theory in Chapter 4. The success or otherwise of directly interconnecting consumer and professional equipment depends to a large extent on how this data is interpreted.

Some professional systems are provided with both consumer and professional interfaces and clearly this offers the ideal solution, but some older equipment had only one electrical interface with a switch to select 'consumer' or 'professional' data characteristics. Often the 'consumer' implementation simply meant that it would accept data with the first bit of channel status set to zero, ignoring most or all of the following data. The similarity in channel status between consumer and professional runs out fairly quickly after the first couple of bits. One of the most common problems is in the signalling of pre-emphasis. When a professional signal with 'emphasis not indicated' is received on a consumer device without any intervention, it will assume that the copy protection flag is being asserted and prevent the signal being recorded (depending on SCMS, as described in section 4.8.7). Similarly, a copy-protected consumer recording being copied directly to a professional system without intervention would force the professional device to a 'no emphasis' state, although a non-copy-protected consumer recording would normally be OK and would set the emphasis state correctly.

Past the emphasis bits there is no similarity at all between the interfaces, and thus it is difficult to say exactly what the results of one system interpreting the other's channel status data would be. Ideally a receiver should check the first bit of channel status to detect the consumer or professional nature of the data, and then switch its implementation automatically to interpret it accordingly.

6.8   Handling differences in audio signal rate and resolution

Because of the variety of sampling rates in use in digital audio and the increasing use of audio sample resolutions beyond 16 bits it is important to ensure that the audio signal retains optimum sound quality when it is converted digitally from one rate or resolution to another. The question of sample rate conversion has already been covered to some extent earlier in this chapter, since it is closely related to the topic of synchronization. There is little more to be said here except to state that normally it is impossible to interconnect two devices digitally whose sampling rates differ by more than a tiny amount from one another, requiring that a sample rate convertor be used between the two. The question of differences in sample word length, though, will be covered in more detail.

The standard two-channel interface allows for up to 24 bits of audio per sample, linearly encoded in fixed point form. Until recently only 16 of these were normally used, with the remaining bits set to zero and the MSB of the 16-bit sample in the bit 27 position, but the question now arises as to how to cope with signals of, say, 18- or 20-bit resolution when they are digitally connected to devices of lower resolution. A number of techniques can be used to process, say, a 20-bit signal to reduce its resolution to 16 bits. These range from straightforward truncation, through bit-shifting and redithering at the new resolution, to developments which involve intelligent rounding of the truncation error by noise shaping. In future it may be that professional digital audio devices will incorporate internal intelligent procedures to handle signals of a higher resolution than their internal architecture allows, but at the moment it is normally necessary to employ external processing of some kind at such a juncture.

Truncation is the worst possible solution and involves simply losing the least significant bits of the word. Without redithering the result of truncation is very unpleasant low-level distortion. If a 20-bit source were connected digitally to a 16-bit destination without any intermediate processing the result would normally be the straightforward truncation of the four LSBs.

The addition of dither noise in the digital domain at the point where resolution is reduced is a suitable means of improving the distortion situation and this has been implemented on some digital interface processors and in professional digital mixers. Some editors also have various dithering algorithms for this purpose. The process randomizes the quantizing error that results from word-length reduction by adding a pseudo-random number sequence of controlled amplitude and spectrum to the incoming audio data. In addition, if the full dynamic range of the digital signal has not been used by the programme material (if headroom has been left, for example) it may be possible to bit-shift the 20-bit samples upwards before truncating and redithering. That way more of the MSBs are used and less of the information contained in the LSBs is lost. This is achieved by a simple increase of gain in the digital domain prior to 16-bit transfer.

In the 1992 revision of the AES3 interface standard, provision is made for much more careful definition of the sample word length, such that receiving devices may optimize the transfer of data from a transmitter of different resolution. Standardization work has also gone on within the EBU to determine how analog signal levels should relate to digital signal levels, especially since 20-bit recording can be used on the audio tracks of digital video recorders. The conclusion reached was that one could not rely on correct implementation of byte 2 of channel status in devices using the AES3 interface in all cases, especially in older equipment. Irrespective of the number of bits, the only practical argument was for a fixed relationship to be used between analog and digital levels[13].

Originally EBU Recommendation R-64 specified that analog alignment level (corresponding to a meter reading of PPM4 or 0 dBu electrically) should be set to read 12 dB below full scale (i.e. −12 dB FS) on a digital system. This was based on the dynamic range available from typical 16-bit convertors, and assumed that the finished programme's level would be well controlled. Since then 16-bit convertor technology has improved and because it was necessary to use the same alignment for 20-bit systems the new recommendation now specifies alignment level to be 18 dB below full scale. This allows for an additional 6 dB of operational headroom.

6.9 Analysing the digital audio interface

Because of the wide variety of implementations possible in the standard interfaces and because of the need to test digital interface signals to determine their 'health' as electrical signals, there have arisen a number of items of test equipment which may be used to analyse the characteristics of the signal. The following is just a short summary of testing techniques, more detailed coverage of which may be found in Cabot[14], Blair *et al.*[15], Mornington West[16], and Stone[17].

6.9.1   Eye pattern and pulse-width testing

The eye pattern of a two-channel interface signal is a rough guide to its electrical 'health' and gives a clue to the likelihood of it being decoded correctly. In a test set-up described by Cabot, pictured in Figure 6.16, it is possible to vary the attenuation and high-frequency roll-off over a digital link so as to 'close the eye' of a random digital audio signal to the limits of the AES3 specification. Having done this one may then verify whether a receiver correctly decodes the data and thus whether it is within the specification. The testing of eye patterns requires a high quality oscilloscope whose trigger input is derived from a stable clock locked to the source clock rate, preferably at a submultiple of it. Alternatively, provided that the clock recovery in the receiver is of high quality and rejects interconnect jitter it may be possible to use this, with the proviso that the eye pattern's reliability will be affected by any instability in the trigger signal.

An alternative to eye pattern testing on an oscilloscope is the use of a stand-alone interface analyser such as that described by Blair *et al.*, capable of displaying the amplitude and 'pulse smear' of the signal on an LCD display with relation to either an internal or external reference signal.

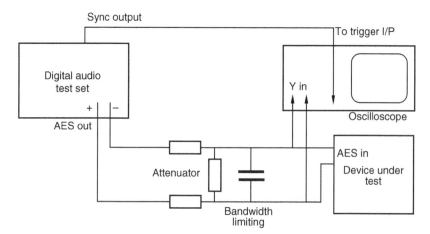

Figure 6.16   This test arrangement is suggested by Cabot for measuring the ability of a device to decode digital input signals at the limits of the eye pattern.

It has been suggested by Kondakor[18] of the BBC Designs and Equipment Department that eye patterns may not always indicate the 'decodability' of a signal, since he has found examples of seemingly good eye patterns which still give problems at the receiver. Consequently a device has been built that measures the variations in pulse width of received signals – a parameter which becomes greater on poor electrical links – displaying the result on a simple visual display which indicates variations in the three possible bit-cell timings (the '1', '0' and preamble wide pulses) in the form of bar lengths. It is claimed that this gives quick and easy verification of the reliability of received signals, and correlates well with the likelihood that a signal will be decoded.

6.9.2   *Security margin estimation*

Often it is necessary to get a rough idea of how close an interface is to failure and one way of doing this is to introduce a broadband attenuator into the link at the receiving end (with sufficient bandwidth to accommodate the digital audio signal). The attenuation is gradually increased until the receiver fails to decode the signal. The amount of attenuation that can be introduced without the link failing is then a rough guide to the margin of security. An alternative to this method is for test equipment to examine the narrow pulses that follow the extra wide pulses in the subframe preambles of the data. Due to intersymbol interference these are usually the first to be reduced in width and level, or even disappear in cases of poor links and limited link bandwidth. Some interface receiver ICs use this criterion as a measure of the security margin in hand.

6.9.3   *Error checking*

Digital interfaces are designed to be used in an error-free environment – in other words, errors are not anticipated over digital links and there is no means of correcting them. This is an achievable situation in well-designed systems and above a certain signal-to-noise ratio one may expect no errors, but below this ratio the error rate will rise quite quickly. The error rate of a digital interface can be checked by a number of means. Common to all the techniques is that a certain bit pattern is generated by the test equipment and the received pattern is then compared against the generated pattern to check for data errors. It is then possible to attempt such things as increasing the noise level on the interface by injecting controlled levels and bandwidths of artificially generated noise to see the effect on error rates. A novel means of checking for errors also described by Cabot is to use a sine-wave test signal and to monitor the digital THD + N (distortion plus noise) reading at the receiver on suitable test equipment. Any error gives rise to a spike in the signal, resulting in a momentary increase in distortion which can be monitored at the output of the notch filter. He points out that errors in LSBs will produce less of an effect on the THD + N reading than errors in MSBs. By correlating errors with the data pattern at the time of the error it is possible to determine whether they are data dependent.

6.9.4 Other tests

Further checks on the interface may be performed on commercial test equipment, such as the accuracy of the sample clock (compared against either an external or calibrated internal reference), measurements of data cell jitter and common mode signal level. Test equipment can also show the states of all the information in channel status and other auxiliary bits if required, many systems converting these into useful human-readable indication. Examples exist of systems that will analyse the incoming channel status and other data and modify it to suit the established requirements of the receiver in order to set up proper communications in problem situations.

6.10 Interface transceiver chips

A number of dedicated ICs are now available for receiving and transmitting standard two-channel interface data, many operating in either consumer or professional modes. Such chips are large-scale integrated circuits (LSIs) which take care of low jitter clock recovery, automatic CRCC checking and generation, automatic sample rate selection and channel status control, among other features. Examples of such devices are the Cirrus Logic (formerly Crystal Semiconductors) series including the CS8427 transceiver chip that will transmit and receive consumer or professional signals at sampling frequencies up to 96 kHz. In such chips the clock recovery is carefully controlled in order to extract a low jitter clock from the interface signal for use by the audio system. The interface between two-channel transceiver chips and the internal signal processing of the equipment is normally in a serial form that can be accepted directly by DSP devices such as the Motorola DSP 56000.

Multichannel interfacing using MADI is based on the FDDI (Fibre Distributed Digital Interface) standard and requires the use of so-called

'TAXI' chips designed by AMD (Advanced Micro Devices). The Am7968 and 7969 chips can be used to handle transmission and reception of the high bit rate data from either optical fibre or copper cables, transferring this data normally in parallel form to and from the internal signal processing of the device in question in order to achieve the high transfer rate necessary.

6.11   Routers and switchers

There are essentially two ways by which standard two-channel interface signals may be routed and switched in systems such as broadcast studio centres where a number of sources and destinations exist. One technique is to use a TDM (Time-Division Multiplexed) switcher, and the other is to use a conventional logic-based crosspoint router. Which is appropriate depends largely on the importance of 'silent' switching, such as might be required for 'hot' or 'on-air' applications, since in a router used simply for assigning signals between sources and destinations (the equivalent to an analog jack-field) it may not matter that discontinuities arise when routing is altered.

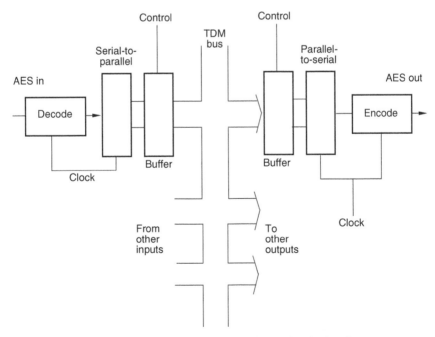

Figure 6.17   In a TDM router, input signals are decoded and multiplexed onto a common high speed bus. Inputs are routed to outputs by extracting data in the appropriate time slot for a particular input channel. Inputs must be synchronous for successful TDM routing.

Inputs

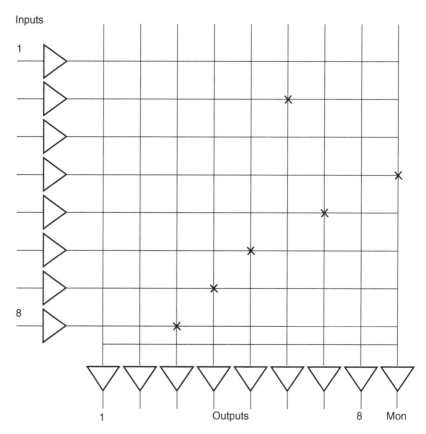

Figure 6.18 A simple crosspoint router may be used to connect a certain input channel to a certain output channel by selecting the appropriate crosspoint. Asynchronous signals may be handled here.

In a TDM switcher all the inputs must be synchronous and must be tightly phased inside the router to allow switching at precise timing points within the AES/EBU frame. The inputs are decoded and time-division multiplexed onto a fast parallel bus (as shown in Figure 6.17), allowing output routing to be determined by extracting data in the appropriate channel time slot on the bus and assigning it to a particular output. As suggested by de Jaham[19], the TDM switcher is appropriate when a large number of sources and destinations are involved because the physical space occupied 'per switching point' is smaller compared with a simple crosspoint router, yet the TDM switcher is complex in design and not usually cost effective for a small to medium number of inputs.

The crosspoint router may be a simple matrix of inputs and outputs, allowing any input to be routed to any output by selecting the appropriate 'crosspoint' between the two, using switching logic (see Figure 6.18). Furthermore, the signals may remain in the serial modulated form. One of the great advantages of the crosspoint router is that the inputs do not need to be synchronous, and it could even handle signals of different sampling rates. The problem with such a router is that the precise timing of the crosspoint may be difficult to control, so a glitch may result in the output signal at the time of switching. Evans[12] suggests that the typical level of such glitches is around −50 dBu in the analog domain, but this of course depends on the relationship between analog and digital signal levels. Roe[20] indicates that clicks caused by such switching are usually masked by the digital filtering used in D/A convertors to a level up to around 50 dB below peak. A reframer may be used to detect and remove the corrupted sample from the data stream where noise-free switching is required. De Jaham also suggests that it is possible to mark the digital frames which have been corrupted at the switching point so that they may be suitably concealed by equipment further down the signal chain, but points out that the process of marking is covered by patents belonging to TDF (Télédiffusion de France).

6.12   Other useful products

6.12.1   Interface format convertors

When operating in a mixed interface format environment, such as when wishing to copy a recording between a consumer and a professional system or when interfacing equipment of different formats, it may be necessary to employ a suitable format convertor. There are a number of such devices on the market, accepting inputs in one of a selection of standard formats (e.g. ADAT, TDIF, SPDIF, AES/EBU) and converting them to one of a selection of output formats. Some more expensive systems also offer synchronization and sample rate conversion so that the system can transfer signals to systems of different sample rates, as well as different formats.

Often interface processors will take care of more subtle but extremely useful functions such as adding correct CRCC information to channel status and checking for correct implementation of the standard. They may also allow the user to alter some or all of the channel status bits manually. With the advent of recording CD machines (CD-R) there have also arisen a number of systems capable of extracting the track start ID information from the user bits of the consumer digital interface, using this information to increment the track ID information of the destination format correctly, such as when a DAT tape is copied to a CD or vice versa. There is also the possibility that the processor may remove SCMS copy protection (see section 4.8.7) to allow professional users the flexibility that was lost with the introduction of the copy management system. Clearly such a device should be used with due regard to copyright laws.

6.12.2   Digital headphones

Once signals are interfaced digitally in a studio centre it becomes more difficult to monitor the signal audibly. In analog systems a pair of high impedance headphones can simply be plugged into a jackfield to determine the presence or lack of a signal, but clearly this would not work in a digital system. There exist a small number of digital headphone products that contain an AES/EBU decoder and a built-in D/A convertor, allowing the user to treat digital signals in a similar way to analog signals.

6.13   A brief troubleshooting guide

If a digital interface between two devices appears not to be working it could be due to one or more of the following conditions. The reader should refer to the main text for more detailed explanation of the conditions described.

Asynchronous sample rates
The two devices must normally operate at the same sampling frequency, preferably locked to a common reference. Ensure that the receiver is in external sync mode and that a synchronizing signal (common to the transmitter) is present at the receiver's sync input. If the incoming signal's transmitter cannot be locked to the reference it must be resynchronized or sample rate converted. Alternatively, set the receiver to genlock to the clock contained in the digital audio input (standard two-channel interfaces only).

The 'Sync' or 'Locked' indicator flashing in or out on the receiver normally means that no sync reference exists or that it is different from that of the signal at the digital input. Check that sync reference and input are at the correct rate and locked to the same source. Decide on whether to use internal or external sync reference, depending on application.

If problems with 'good lock' or drifting offset arise when locking to other machines or when editing, check that any timecode is synchronous with the video and sampling rate. If not, the tape must be restriped with timecode locked to the same reference as the recorder, or a synchronizer used which will lock a digital audio input to the rate dictated by a timecode input.

Sampling frequency mode

The transmitter may be operating in the AES3 single-channel-double-sampling-frequency mode in which case successive subframes will carry adjacent samples of a single channel at twice the normal sampling frequency. This might sound like audio pitch-shifted downwards if decoded and converted by a standard receiver incapable of recognizing this mode. Alternatively the devices may be operating at entirely different sampling frequencies and therefore not communicating.

Digital input

It may be that the receiver is not switched to accept a digital input.

Data format

Received data is in the wrong format. Both transmitter and receiver must operate to the same format. Conflicts may exist in such areas as channel status, and there may be a consumer–professional conflict. Use a format convertor to set the necessary flags.

Non-audio or 'other uses' set

The data transmitted over the interface may be data-reduced audio, such as AC-3 or DTS format. It can only be decoded by receivers specially designed for the task. The data will sound like noise if it is decoded and converted by a standard linear PCM receiver, but in such receivers it will normally be muted because of the indication in channel status and/or the validity bit.

Cables and connectors

Cables or connectors may be damaged or incorrectly wired. The cable may be too long, of the wrong impedance, or generally of poor quality. The digital signal may be of poor quality. Check eye height on the scope against specification and check for possible noise and interference sources. Alternatively make use of an interface analyser.

SCMS (consumer interface only)

The copy protect or SCMS flag may be set by the transmitter. For professional purposes, use a format convertor to set the necessary flags or use the professional interface which is not subject to SCMS.

Receiver mode

The receiver is not in record or input monitor mode. Some recorders must be at least in record–pause before they will give an audible and metered output derived from a digital input.

References

1. AES, *AES11-1997. Synchronization of digital audio equipment in studio operations* (1997)
2. AES, AES5-1984. AES recommended practice for professional digital audio applications employing pulse code modulation – preferred sampling frequencies. *Journal of the Audio Engineering Society*, vol. 32, pp. 781–785 (1984)
3. Shelton, W.T., Synchronization of digital audio. In *Proceedings of the AES/EBU Interface Conference*, 12–13 September, London, pp. 92–116, Audio Engineering Society British Section (1989)
4. Dunn, N.J., Jitter: specification and assessment in digital audio equipment. Presented at the *93rd AES Convention, San Francisco*, 1–4 October, preprint no. 3361 (C-2) (1992)
5. Dunn, N.J., Considerations for interfacing digital audio equipment to the standards AES3, AES5 and AES11. In *Proceedings of the AES 10th International Conference*, 7–9 September, pp. 115–126 (1991)
6. Dunn, C. and Hawksford, M.O., Is the AES/EBU/SPDIF digital audio interface flawed? Presented at the *93rd AES Convention, San Francisco*, 1–4 October, preprint no. 3360 (C-1) (1992)
7. Gilchrist, N.H.C., Sampling-rate synchronization of digital sound signals by variable delay. *EBU Technical Review*, no. 183, October (1980)
8. Parker, M., Sample frequency conversion, sample slippage, pitch changing and varispeed. In *Proceedings of the AES 10th International Conference*, 7–9 September, p. T-69 (1991)
9. Shelton, W.T., Timing inter-relations for audio with video. In *Proceedings of the AES 9th International Conference*, 1–2 February, pp. 31–44 (1991)
10. Bensberg, G., Time for digital audio within television. Presented at the *92nd AES Convention, Vienna*, 24–27 March, preprint no. 3258 (1992)
11. Komly, A. and Viallevielle, A., *Synchronization and time codes in the user channel*. Proposal submitted to AES SC 2-5-1 working party on synchronization, Paris, March (1990)
12. Evans, P., Digital audio in the broadcast centre. *EBU Technical Review*, no. 241/242, June–August (1990)
13. Møller, L., Signal levels across the EBU/AES digital audio interface. In *Proceedings of the 1st NAB Radio Montreux Symposium, Montreux, Switzerland*, 10–13 June, pp. 16–28, National Association of Broadcasters (1992)
14. Cabot, R., Measuring AES/EBU digital audio interfaces. *Journal of the Audio Engineering Society*, vol. 38, no. 6, October, pp. 459–467 (1989)
15. Blair, I. *et al.*, New techniques in analysing the digital audio interface. Presented at the *92nd AES Convention, Vienna*, 24–27 March, preprint no. 3230 (1992)
16. Mornington West, A., Signal analysis. In *Proceedings of the AES/EBU Interface Conference*, 12–13 September, pp. 83–91, Audio Engineering Society British Section (1989)
17. Stone, D., Digital signal analysis. *International Broadcast Engineer*, November (1992)
18. Kondakor, K., A pulse width analyser for the rapid testing of the AES/EBU serial digital audio signals. Presented at the *93rd AES Convention, San Francisco*, 1–4 October (1992)
19. De Jaham, S., Asynchronous routing. In *Proceedings of the AES/EBU Interface Conference*, 12–13 September, pp. 64–68, Audio Engineering Society British Section (1989)
20. Roe, G., Integrated routing in a hybrid environment. *International Broadcast Engineer*, July, pp. 44–46 (1991)

7

Digital video interfaces

In this chapter the various standardized interfaces for component and composite video will be detailed along with the necessary troubleshooting techniques.

7.1   Introduction

Of all the advantages of digital video, the most important for production work is the ability to pass through multiple generations without quality loss. Digital interconnection between such production equipment is highly desirable to avoid the degradation due to repeated conversions.

Video convertors universally use parallel connection, where all bits of the pixel value are applied simultaneously to separate pins. Disk drives lay data serially on the track, but within the circuitry, parallel presentation is more common because it allows slower, and hence cheaper, memory chips to be used for timebase correction. Reed–Solomon error correction depends upon symbols assembled from typically eight bits. Digital effects machines and switchers typically operate upon pixel values in parallel.

The first digital video interfaces were based on parallel transmission. All that is necessary is a set of suitable driver chips, running at an appropriate sampling rate, to send video data down cables having separate conductors for each bit of the sample, along with a clock to tell the receiver when to sample the bit values. The complexity is trivial, and for short distances this approach represented the optimum solution at the time.

Parallel connection has drawbacks too; these come into play when longer distances are contemplated. A multicore cable is expensive, and the connectors are physically large. It is difficult to provide good screening of a multicore cable without it becoming inflexible. More seriously, there are electronic problems with multicore cables. The propagation speeds of pulses down all of the cores in the cable will not be exactly the same, and so, at the end of a long cable, some bits may still be in transition when the clock arrives, whilst others may have begun to change to the value in the next pixel.

Where it is proposed to interconnect a large number of units with a router, that device will be extremely complex because of the number of parallel signals to be handled. In short, parallel technology could not and did not replace the central analog router of a conventional television station. The answer to these problems is the serial connection. All of the digital samples are multiplexed into a serial bitstream, and this is encoded to form a self-clocking channel code which can be sent down a single channel. Skew caused by differences in propagation speed cannot then occur. The bit rate necessary is in excess of 200 Mbits/s for standard definition and almost 1.5 Gbits/s for HD, but this is well within the capabilities of coaxial cable.

The cabling savings implicit in serial systems are obvious, but the electronic complexity of a serial interconnect is naturally greater, as high speed multiplexers or shift registers are necessary at the transmitting end, and a phase-locked loop, data separator and deserializer, as outlined in Chapter 3, are needed at the receiver to regenerate the parallel signal needed within the equipment. The availability of specialized serial chips from a variety of manufacturers meant that serial digital video would render parallel interfaces obsolete very quickly.

A distinct advantage of serial transmission is that a matrix distribution unit or router is more easily realized. Where numerous pieces of video equipment need to be interconnected in various ways for different purposes, a crosspoint matrix is an obvious solution. With serial signals, only one switching element per signal is needed. A serial system has a potential disadvantage that the time distribution of bits within the block has to be closely defined, and, once standardized, it is extremely difficult to increase the word length if this is found to be necessary. The serial digital interface (SDI) was designed from the outset for 10-bit working but incorporates two low-order bits that may be transmitted as zero in eight-bit applications. In a parallel interconnect, the word extension can be achieved by adding extra conductors alongside the existing bits, which is much easier.

The third interconnect to be considered uses fibre optics. The advantages of this technology are numerous: the bandwidth of an optical fibre is staggering, and is practically limited only by the response speed of the light source and sensor, and for this reason it has been adopted for digital HDTV interfacing. The optical transmission is immune to electromagnetic interference from other sources, nor does it contribute any. This is advantageous for connections between cameras and control units, where a long cable run may be required in outside broadcast applications. The cable can be made completely from insulating materials, so that ground loops cannot occur, although many practical fibre-optic cables include electrical conductors for power and steel strands for mechanical strength.

Drawbacks of fibre optics are few. They do not like too many connectors in a given channel, as the losses at a connection are much greater than with an electrical plug and socket. It is preferable for the only breaks in the fibre to be at the transmitting and receiving points. For similar reasons, fibre optics are less suitable for distribution, where one source feeds many destinations. The bi-directional open-collector or tri-state buses of electronic systems cannot be implemented with fibre optics, nor is it easy to build a crosspoint matrix.

The high frequencies involved in digital video mean that accurate signal termination of electrical cables is mandatory. Cable losses cause the signal amplitude to fall with distance. As a result the familiar passive loop-through connection of analog video is just not possible. Whilst much digital equipment appears to have loop-through connections, close examination will reveal an amplifier symbol joining the input and output. Digital equipment must use active loop-through and if a unit loses power, the loop-through output will fail.

7.2 Areas of standardization

For some time digital interface standards have existed for 525/59.94 and 625/50 4:2:2 component and $4F_{SC}$ composite. More recently digital interface standards for a variety of HD scanning standards have been set.

Digital interfaces require to be standardized in the following areas: connectors, to ensure plugs mate with sockets; pinouts; electrical signal specification, to ensure that the correct voltages and timing are transferred; and protocol, to ensure that the meaning of the data words conveyed is the same to both devices. As digital video of any type is only data, it follows that the same physical and electrical standards can be used for a variety of protocols. It also follows that the same protocols can be conveyed down a variety of physical channels. Figure 7.1 shows that serial, parallel and optical fibre interfaces may carry exactly the same data.

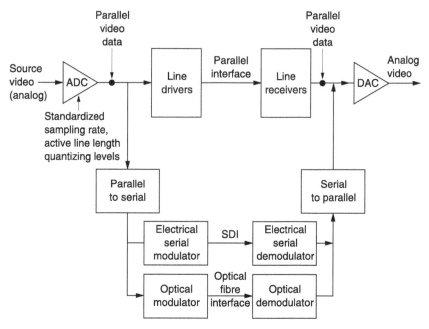

Figure 7.1 If a video signal is digitized in a standardized way, the resulting data may be electrically transmitted in parallel or serial or sent over an optical fibre.

Parallel interfaces are obsolete, but they are worthy of study because the parallel interface standard actually contains a comprehensive definition of how a television signal is described in the digital domain as a series of binary numbers. This definition will include such factors as how the image is described as a pixel array, the sampling rate needed to do that, how the colour is subsampled, the colorimetry and gamma assumed, the connection between analog signal voltage and the value of binary codes and so on. The actual parallel interface usually consists of little more than a number of ECL line drivers and receivers, one for each bit along with a clock. A serial interface will require some form of shift register to convert the word format data into a bitstream along with a channel coder to turn the bitstream into a self-clocking waveform compatible with the channel, which may be electrical or optical. The SD and HD serial interfaces are quite similar except, of course, for the bit rate.

7.3 Digitizing component video

It is not necessary to digitize analog sync pulses in component systems, since the only useful video data is that sampled during the active line.

As the sampling rate is derived from sync, it is only necessary to standardize the size and position of a digital active line and all other parts of the video waveform can be recreated at a later time. The position is specified as a given number of sampling clock periods from the leading edge of sync, and the length is simply a standard number of samples. The digital active line is somewhat longer than the analog active line to allow for some drift in the line position of the analog input and to place edge effects in digital filters outside the screen area. Some of the first and last samples of the digital active line will represent blanking level, thereby avoiding abrupt signal transitions caused by a change from blanking level to active signal. When converting analog signals to digital it is important that the analog unblanked picture should be correctly positioned within the line. In this way the analog line will be symmetrically disposed within the digital active line. If this is the case, when converting the data back to the analog domain, no additional blanking will be necessary, as the blanking at the ends of the original analog line will be recreated from the data. The DAC can pass the whole of the digital active line for conversion and the result will be a correctly timed analog line with blanking edges in the right position.

However, if the original analog timing was incorrect, the unblanked analog line may be too long or off-centre in the digital active line. In this case a DAC may apply digital blanking to the line data prior to conversion. Some equipment gives the user the choice of using blanking in the data or locally applied blanking prior to conversion.

In addition to specifying the location of the samples, it is also necessary to standardize the relationship between the absolute analog voltage of the waveform and the digital code value used to express it so that all machines will interpret the numerical data in the same way. These relationships are in the voltage domain and are independent of the scanning standard used. Thus the same relationships will be found in both SD and HD component formats. As a digital interface is just an alternative way of sending a television picture, the information it contains about that picture will be the same. Thus digital interfaces assume the same standards for gamma and colour primaries as the original analog system.

Figure 7.2 shows how the luminance signal fits into the quantizing range of a digital system. Numbering for 10-bit systems is shown with figures for eight bits in brackets. Black is at a level of 64_{10} (16_{10}) and peak white is at 940_{10} (235_{10}) so that there is some tolerance of imperfect analog signals and overshoots caused by filter ringing. The sync pulse will clearly go outside the quantizing range, but this is of no consequence as conventional syncs are not transmitted. The visible voltage range fills the quantizing range and this gives the best possible resolution.

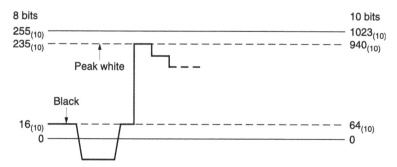

Figure 7.2 The standard luminance signal fits into eight- or ten-bit quantizing structures as shown here.

The colour difference signals use offset binary, where 512_{10} (128_{10}) is the equivalent of blanking voltage. The peak analog limits are reached at 64_{10} (16_{10}) and 960_{10} (240_{10}) respectively, allowing once more some latitude for maladjusted analog inputs and filter ringing.

Note that the code values corresponding to all ones or all zeros (i.e. the two extreme ends of the quantizing range) are not allowed to occur in the active line as they are reserved for synchronizing. ADCs must be followed by circuitry that detects these values and forces the code to the nearest legal value if out-of-range analog inputs are applied. Processing circuits that can generate these values must employ digital clamp circuits to remove the values from the signal. Fortunately this is a trivial operation.

The peak-to-peak amplitude of Y is 880 (220) quantizing intervals, whereas for the colour difference signals it is 900 (225) intervals. There is thus a small gain difference between the signals. This will be cancelled out by the opposing gain difference at any future DAC, but must be borne in mind when digitally converting to other standards.

The sampling rate used in SD was easily obtained as only two scanning standards had to be accommodated. It will be seen that in HD there are further constraints. In principle, the sampling rate of a system need only satisfy the requirements of sampling theory and filter design. Any rate that does so can be used to convey a video signal from one place to another. In practice, however, there are a number of factors that limit the choice of sampling rate considerably.

It should be borne in mind that a video signal represents a series of two-dimensional images. If a video signal is sampled at an arbitrary frequency, samples in successive lines and pictures could be in different places. If, however, the video signal is sampled at a rate which is a multiple of line rate the result will be that samples on successive lines will be in the same place and the picture will be converted to a neat array having vertical columns of samples that are in the same place in all pictures. This allows for the spatial and temporal processing needed in, for example, standards convertors and MPEG coders. A line-locked sampling rate can conveniently be obtained by multiplication of the H-sync frequency in a phase-locked loop. The position of samples along the line is then determined by the leading edge of sync.

Considering SD sampling rates first, whilst the bandwidth required by 525/59.94 is less than that required by 625/50, and a lower sampling rate might have been used, practicality suggested a common sampling rate. The benefit of a standard H-locked sampling rate for component video is that the design of standards convertors is simplified and DVTRs have a constant data rate independent of standard. This was the goal of CCIR (now ITU) Recommendation 601[1], which combined the 625/50 input of EBU Doc. Tech. 3246 and 3247 with the 525/59.94 input of SMPTE RP 125.

CCIR 601 recommends the use of certain sampling rates which are based on integer multiples of the carefully chosen fundamental frequency of 3.375 MHz. This frequency is normalized to 1 in the document.

In order to sample 625/50 luminance signals without quality loss, the lowest multiple possible is 4 which represents a sampling rate of 13.5 MHz. This frequency line-locks to give 858 sample periods per line in 525/59.94 and 864 sample periods per line in 625/50. The spectra of such sampled luminance are shown in Figure 7.3.

In the component analog domain, the colour difference signals typically have one-half the bandwidth of the luminance signal. Thus a sampling rate multiple of 2 is used and results in 6.75 MHz. This sampling rate allows respectively 429 and 432 sample periods per line.

Component video sampled in this way has a 4:2:2 format. Whilst other combinations are possible, 4:2:2 is the format for which the majority of

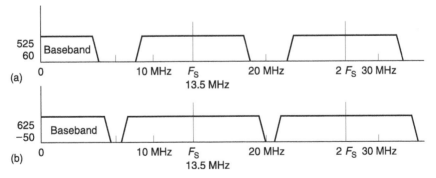

Figure 7.3 Spectra of video sampled at 13.5 MHz. In (a) the baseband 525/60 signal at left becomes the sidebands of the sampling rate and its harmonics. In (b) the same process for the 625/50 signal results in a smaller gap between baseband and sideband because of the wider bandwidth of the 625 system. The same sampling rate for both standards results in a great deal of commonality between 50 Hz and 60 Hz equipment.

production equipment is constructed and is the only SD component format for which parallel and serial interface standards exist. Figure 7.4 shows the spatial arrangement given by 4:2:2 sampling. Luminance samples appear at half the spacing of colour difference samples, and every other luminance sample is co-sited with a pair of colour difference samples. Co-siting is important because it allows all attributes of one picture point to be conveyed with a three-sample vector quantity. Modification of the three samples allows such techniques as colour correction to be performed. This would be difficult without co-sited information. Co-siting is achieved by clocking the three ADCs simultaneously. In some equipment one ADC is multiplexed between the two colour difference signals. In order to obtain co-sited data it will then be necessary to have an analog delay in one of the signals.

For full bandwidth RGB working, 4:4:4 can be used with a possible 4:4:4:4 used if including a key signal. For lower bandwidths, multiples of 1 and 3 can also be used for colour difference and luminance respectively. 4:1:1 delivers colour bandwidth in excess of that required by the composite formats. 4:1:1 is used in the 525 line version of the DVC quarter-inch digital video format. 3:1:1 meets 525 line bandwidth requirements. The factors of 3 and 1 do not, however, offer a columnar structure and are inappropriate for quality post-production.

In 4:2:2 the colour difference signals are sampled horizontally at half the luminance sampling rate, yet the vertical colour difference sampling rates are the same as for luminance. Where bandwidth is important, it is

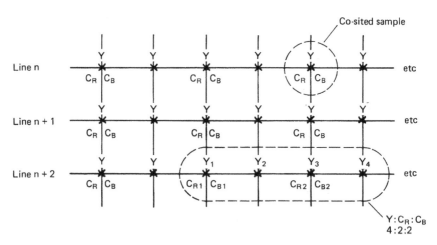

Figure 7.4 In CCIR-601 sampling mode 4:2:2, the line synchronous sampling rate of 13.5 MHz results in samples having the same position in successive lines, so that vertical columns are generated. The sampling rates of the colour difference signals C_R, C_B are one-half of that of luminance, i.e. 6.75 MHz, so that there are alternate Y only samples and co-sited samples which describe Y, C_R and C_B. In a run of four samples, there will be four Y samples, two C_R samples and two C_B samples, hence 4:2:2.

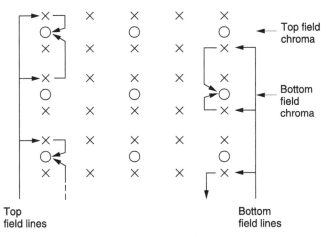

Figure 7.5 In 4:2:0 coding the colour difference pixels are downsampled vertically as well as horizontally. Note that the sample sites need to be vertically interpolated so that when two interlaced fields are combined the spacing is even.

possible to halve the vertical sampling rate of the colour difference signals as well. Figure 7.5 shows that in 4:2:0 sampling, the colour difference samples only exist on alternate lines so that the same vertical and horizontal resolution is obtained. 4:2:0 is used in the 625 line version of the DVC format and in the MPEG 'Main Level Main Profile' format for multimedia communications and, in particular, DVB.

Figure 7.6 shows that in 4:2:2 there is one luminance signal sampled at 13.5 MHz and two colour difference signals sampled at 6.75 MHz. Three separate signals with different clock rates are inconvenient and so multiplexing can be used. If the colour difference signals are multiplexed into one channel, then two 13.5 MHz channels will be required. Such an approach is commonly found within digital component processing equipment where the colour difference processing can take place in a single multiplexed channel.

If the colour difference and luminance channels are multiplexed into one, a 27 MHz clock will be required. The word order is standardized to be:

$$C_b, Y, C_r, Y, \text{ etc.}$$

In order unambiguously to demultiplex the samples, the first sample in the line is defined as C_b and a unique sync pattern is required to identify the beginning of the multiplex sequence. HD adopts the same principle but the frequencies are higher.

There are two ways of handling 16:9 aspect ratio video in SD. In the anamorphic approach both the camera and display scan wider but there is no change to the sampling rates employed and therefore the same

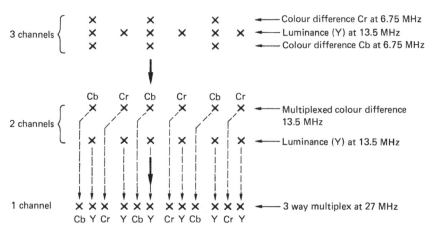

Figure 7.6 The colour difference sampling rate is one-half that of luminance, but there are *two* colour difference signals, C_r and C_b, hence the colour difference data rate is equal to the luminance data rate, and a 27 MHz interleaved format is possible in a single channel.

27 MHz data stream can be employed unchanged. Compared with 4:3, the horizontal spacing of the pixels in 16:9 must be greater as the same number are stretched across a wider picture. This results in a reduction of horizontal resolution, but standard 4:3 production equipment can be used subject to some modifications to the shape of pattern wipes in vision mixers. When viewed on a 4:3 monitor anamorphic signals appear squeezed horizontally.

In the second approach, the pixel spacing is kept the same as in 4:3 and the number of samples per active line must then be increased by 16:12. This requires the data rate to rise to 36 MHz. Thus the luminance sampling rate becomes 18 MHz and the colour difference sampling rate becomes 9 MHz. Strictly speaking the format no longer adheres to CCIR-601 because the sampling rates are no longer integer multiples of 3.375 MHz. If, however, 18 MHz is considered to be covered by Rec. 601, then it must be described as 5.333 ... : 2.666 ... : 2.666....

If the sampling rate is chosen to be a common multiple of the US and European line rates, the spacing between the pixels that results will have to be accepted. In computer graphics, pixels are always square, which means the horizontal and vertical spacing is the same. In 601 sampling, the pixels are not square and their aspect ratio differs between the US and European standards. This is because the horizontal sampling rate is the same but the number of lines in the picture is different.

When CCIR 601 was being formulated, the computer and television industries were still substantially separate and the lack of square pixels was not seen as an issue. In 1990 CCIR 709 recommended that HD formats should be based on 1920 pixels per active line, and use sampling rates based on 2.25 MHz (6.75/3): an unnecessarily inflexible approach again making it unlikely that square pixels would result at all frame rates.

Subsequently, the convergence of computer, film and television technology has led to square pixels being adopted in HD formats at all frame rates, a common sampling rate having been abandoned. Another subtle change is in the way of counting lines. In traditional analog video formats, the number of lines was the total number, including blanking, whereas in computers the number of lines has always been the number visible on the screen, i.e. the height of the pixel array. HD standards adopted the same approach. Thus in the 625 line standard, there will be 625 line periods per frame. Whereas in the 1080 line HD standard there are actually 1125 line periods per frame.

It is slowly being understood that improved picture quality comes not from putting more pixels into the image but from eliminating interlace and increasing the frame rate[2]. Thus 1280×720 progressively scanned frames are described in SMPTE 296M[3]. Unfortunately there are still those who believe that data describing digital television images somehow differs from computer data. The bizarre adherence to the obsolete principle of interlacing seems increasingly to be based on maintaining an artificial difference between computers and television for marketing purposes rather than on any physics or psycho-optics. The failure of the ATSC and FCC to understand these principles has led to a damaging proliferation of HD television standards, in which the simple and effective approach of digital standard definition has been lost.

7.4 Structure of SD component digital

The sampling rate for luma is H-synchronous 13.5 MHz. This is divided by two to obtain the colour difference sampling rate. Figure 7.7 shows that in 625 line systems the control system[4] waits for 132 luma sample periods after an analog sync edge before commencing sampling the line. Then 720 luma samples and 360 of each type of colour difference sample are taken; 1440 samples in all. A further 12 sample periods will elapse before the next sync edge, making $132 + 720 + 12 = 864$ sample periods. In 525 line systems[5], the analog active line is in a slightly different place and so the controller waits 122 sample periods before taking the same digital active line samples as before. There will then be 16 sample periods before the next sync edge, making $122 + 720 + 16 = 858$ sample periods.

For 16:9 aspect ratio working, the line and field rate remain the same, but the luminance sampling rate may be raised to 18 MHz and the colour difference sampling rates are raised to 9 MHz. This results in the sampling

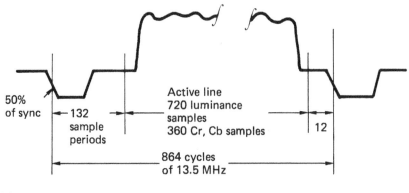

(a)

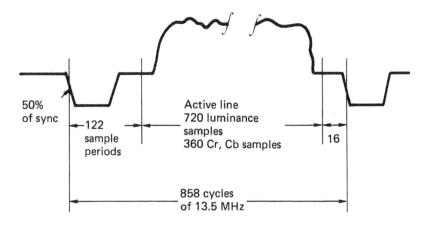

(b)

Figure 7.7 (a) In 625 line systems to CCIR-601, with 4:2:2 sampling, the sampling rate is exactly 864 times line rate, but only the active line is sampled, 132 sample periods after sync. (b) In 525 line systems to CCIR-601, with 4:2:2 sampling, the sampling rate is exactly 858 times line rate, but only the active line is sampled, 122 sample periods after sync. Note active line contains exactly the same quantity of data as for 50 Hz systems.

structure shown for 625 lines in Figure 7.8(a) and for 525 lines in (b). There are now 960 luminance pixels and 2×480 colour difference pixels per active line.

7.5 Structure of HD component digital

Given the large number of HD scanning standards, it is only possible to outline the common principles here. Specific standards will differ in line and sample counts. Those who are accustomed to analog SD will note

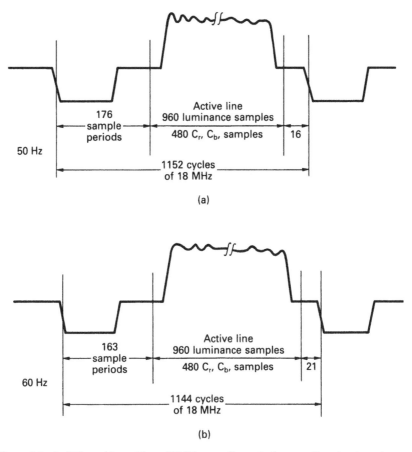

Figure 7.8 In 16:9 working with an 18 MHz sampling rate the sampling structure shown here results.

that in HD the analog sync pulses are different. In HD, the picture quality is more sensitive to horizontal scanning jitter and so the signal-to-noise ratio of the analog sync edge is improved by doubling the amplitude. Thus the sync edge starts at the most negative part of the waveform, but continues rising until it is as far above blanking as it was below. As a result 50% of sync, the level at which slicing of the sync pulse is defined to take place, is actually at blanking level. All other voltages and gamuts remain the same as for SD.

The treatment of SD formats introduced the concept of the digital active line being longer than the analog line. Some HD formats have formalized this by describing the total active pixel array as the production aperture, and the slightly smaller area within that, corresponding to the unblanked area of the analog format, as the clean aperture. The quantizing standards of HD are the same as for SD, except that the option of 12-bit resolution is added.

SMPTE 274M[6] describes 1125 lines per frame 16:9 aspect ratio HD standards having a production aperture of 1920×1080 pixels and a clean aperture of 1888×1062 pixels. The standard uses square pixels, thus $1080 \times 16 = 1920 \times 9$. Both interlaced and progressive scanning are supported, at a wide variety of frame rates: basically 24, 25, 30, 50 and 60 Hz with the option of incorporating the reduction in frequency of 0.1% for synchronization to the traditional NTSC timing.

As with SD, the sampling clock is line locked. However, there are some significant differences between the SD and HD approaches. In SD, a common sampling rate is used for both line standards. This allows both a common interface data rate and an interface that works in real time, but results in pixels that are not square. In HD, the pixels are square and this causes the video sampling rate to change with the frame rate. In order to keep the interface bit rate constant, variable amounts of packing are placed between the active lines but the result is that the interface no longer works in real time at all frame rates and requires buffering at source and destination. The interface symbol rate has been chosen to be a common multiple of 24, 25 and 30 times 1125 Hz so that there can always be an integer number of interface symbol periods in a line period.

For example, if used at 30 Hz frame rate interlaced, there would be $1125 \times 30 = 33\,750$ lines per second. Figure 7.9(a) shows that the luma sampling rate is 74.25 MHz and there are 2200 cycles of this clock in one line period. From these, 1920 cycles correspond to the active line and 280 remain for blanking and TRS. The colour difference sampling rate is one half that of luma at 37.125 MHz and 960 cycles correspond to the active line. As there are two colour difference signals, when multiplexed together the symbol rate will be $74.25 + 37.125 + 37.125 = 148.5$ MHz. The standard erroneously calls this the interface sampling rate, which is not a sampling rate at all, but a word rate or symbol rate.

Thus the parallel interface has a clock rate of 148.5 MHz. When ten-bit symbols are serialized, the bit rate becomes 1.485 GHz, the bit rate of serial HD. If the option of adhering to the picture rate reduction of 0.1% is taken, all of the above frequencies fall by that amount.

If the frame rate is reduced to 25 Hz, as in (b), the line rate falls to $1125 \times 25 = 28\,125$ Hz and the luma sampling rate falls to $2200 \times 28\,125 = 61.875$ MHz. The interface symbol rate does not change, but remains at 148.5 MHz. In order to carry 50 Hz pictures, time compression is used. At 28 125 lines per second, there will be 2640 cycles of 74.25 MHz, the luma interface rate, per line, rather than the 2200 cycles obtained at 60 Hz. Thus the line still contains 1920 active luma samples, but for transmission, the number of blanking/TRS cycles has been increased to 720.

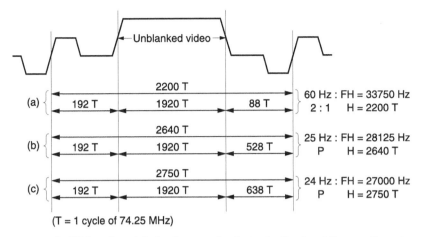

Figure 7.9 In HD interfaces it is the data rate that is standardized, not the sampling rate. At (a) an 1125/30 picture requires a luma sampling rate of 74.25 MHz to have 1920 square pixels per active line. The data rate of the chroma is the same, thus the interface symbol rate is 148.5 MHz. At (b) with 25 Hz pictures, the symbol rate does not change. Instead the blanking area is extended so the data rate is maintained by sending more blanking. At (c) an extension of this process allows 24 Hz material to be sent.

Although the luma is sampled at 61.875 MHz, for transmission luma samples are placed in a buffer and read out at 74.25 MHz. This means that the active line is sent in less than an active line period.

Figure 7.9(c) shows that a similar approach is taken with 24 Hz material in which the number of blanking cycles is further increased.

The 1.485 GHz rate is adequate for interlaced video and for progressively scanned film, in which the frame rate is only 24 or 25 Hz. However, for progressively scanned video, the frame rate may be as high as 60 Hz and this would require the bit rate to be doubled. However, it is becoming increasingly known that because progressive scan eliminates interlace artefacts, it does not need twice the data rate of interlaced systems to give the same perceived quality. Resolution falls dramatically in the presence of even quite slow motion in interlaced video, whereas in progressively scanned video it does not[2]. Thus on real moving pictures, progressively scanned systems with relatively modest static resolution give better performance because that resolution is maintained. Consequently the 50 and 60 Hz 1920 × 1080 progressive standards are quite unnecessary for television purposes.

SMPTE 296M describes the 720P standard[3] that gives the best results of all of the television industry standards, although not as good as the progressive standard developed by the US military which has a higher frame rate.

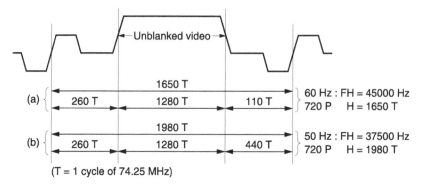

Figure 7.10 In progressively scanned images, the level of artefacts is much lower than interlaced systems allowing the quality to be maintained with fewer picture lines. At (a) the structure of 720P/60 is shown to have the same data rate as 1080I/30. Lower frame rates are supported by blanking extension as shown in (b).

720P uses frames containing 750 lines of which 30 correspond to the vertical interval. Note that as interlace is not used, the number of lines per frame does not need to be odd. 720P has square pixels and so must have $720 \times 16/9 = 1280$ pixels per line. The production aperture is thus 1280×720 pixels. A clean aperture is not defined.

The 1280×720 frame can be repeated at 60, 50, 30, 25 and 24 Hz. The same interface symbol rate as 274M is used, so clearly this must also be a common multiple of 24, 25, 30, 50 and 60 times 750 Hz.

Figure 7.10(a) shows that 720/60 has a line rate of 45 kHz and has 1650 sample periods per line, corresponding to a luma sampling rate of 74.25 MHz. The colour difference sampling rate is half of that, but as there are two colour difference signals, the overall symbol rate becomes 148.5 MHz and so this format transmits in real time. As the frame rate goes down, the number of interface symbol periods per line will rise, but the number of pixels remains constant at 1280. As a result the number of blanked symbols rises, as does the degree of time compression of the transmission of each line. This is shown in the remainder of Figure 7.10.

7.6 Synchronizing

The component interface carries a multiplex of luminance and colour difference samples and it is necessary to synchronize the demultiplexing process at the receiver so that the components are not inadvertently transposed. As conventional analog syncs are discarded, horizontal and vertical synchronizing must also be provided. In the case of serial transmission it is also necessary to identify the position of word boundaries so that correct deserialization can take place. These functions are performed by

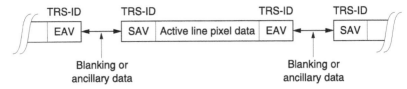

Figure 7.11 The active line data is bracketed by TRS-ID codes called SAV and EAV.

special bit patterns known as timing reference and identification signals (TRS-ID) sent with each line. TRS-ID differs only slightly between formats. Figure 7.11 shows the location of TRS-ID. Immediately before the digital active line location is the SAV (start of active video) TRS-ID pattern, and immediately after is the EAV (end of active video) TRS-ID pattern. These unique patterns occur on every line and continue throughout the vertical interval.

Each TRS-ID pattern consists of four symbols: the same length as the component multiplex repeating structure. In this way the presence of a TRS-ID does not alter the phase of the multiplex. Three of the symbols form a sync pattern for deserializing and demultiplexing (TRS) and one is an identification symbol (ID) that replaces the analog sync signals. The first symbol contains all ones and the next two contain all zeros. This bit sequence cannot occur in active video, even due to concatenation of successive pixel values, so its detection is reliable. As the transition from a string of ones to a string of zeros occurs at a symbol boundary, it is sufficient to enable unambiguous deserialization, location of the ID symbol and demultiplexing of the components. Whatever the word length of the system, all bits should be either ones or zeros during TRS.

The fourth symbol in the ID contains three data bits, H, F and V. These bits are protected by four redundancy bits which form a seven-bit Hamming codeword.

Figure 7.12(a) shows how the Hamming code is generated. Single bit errors can be corrected and double bit errors can be detected according to the decoding table in (b).

Figure 7.13(a) shows the structure of the TRS-ID. The data bits have the following meanings:

- H is used to distinguish between SAV, where it is set to 0, and EAV where it is set to 1.
- F defines the state of interlace and is 0 during the first field and 1 during the second field. F is only allowed to change at EAV. In interlaced systems, one field begins at the centre of a line, but there is no sync pattern at that location so the field bit changes at the end of the line in which the change took place.

Bit	9	8 F	7 V	6 H	5 P3	4 P2	3 P1	2 P0	1	0
	1	0	0	0	0	0	0	0	0	0
	1	0	0	1	1	1	0	1	0	0
	1	0	1	0	1	0	1	1	0	0
	1	0	1	1	0	1	1	0	0	0
	1	1	0	0	0	1	1	1	0	0
	1	1	0	1	1	0	1	0	0	0
	1	1	1	0	1	1	0	0	0	0
	1	1	1	1	0	0	0	1	0	0

Data   Check bits

(a)

Received P3 – P0	Received bits 8, 7, and 6 (F, V, and H)							
	000	001	010	011	100	101	110	111
0000	000	000	000	*	000	*	*	111
0001	000	*	*	111	*	111	111	111
0010	000	*	*	011	*	101	*	*
0011	*	*	010	*	100	*	*	111
0100	000	*	*	011	*	*	110	*
0101	*	001	*	*	100	*	*	111
0110	*	011	011	011	100	*	*	011
0111	100	*	*	011	100	100	100	*
1000	000	*	*	*	*	101	110	*
1001	*	001	010	*	*	*	*	111
1010	*	101	010	*	101	101	*	101
1011	010	*	010	010	*	101	010	*
1100	*	001	110	*	110	*	110	110
1101	001	001	*	001	*	001	010	*
1110	*	*	*	011	*	101	110	*
1111	*	001	010	*	100	*	*	*

(b)

Figure 7.12 The data bits in the TRS are protected with a Hamming code which is calculated according to the table in (a). Received errors are corrected according to the table in (b) where a dot shows an error detected but not correctable.

- V is 1 during vertical blanking and 0 during the active part of the field. It can only change at EAV.

Figure 7.13(b) (top) shows the relationship between the sync pattern bits and 625 line analog timing, whilst below is the relationship for 525 lines.

Figure 7.14 shows a decode table for SD TRS which is useful when interpreting logic analyser displays.

The same TRS-ID structure is used in SMPTE 274M and 296M HD. It differs in that the HD formats can support progressive scan in which the F bit is always set to zero.

7.7 Component ancillary data

In component standards, only the active line is transmitted and this leaves a good deal of spare capacity. The two line standards differ on how

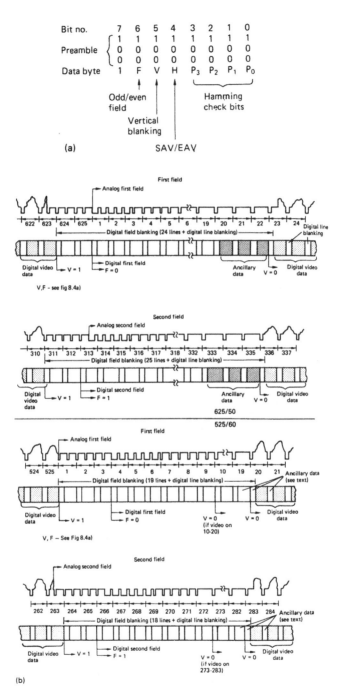

Figure 7.13 (a) The four-byte synchronizing pattern which precedes and follows every active line sample block has this structure. (b) The relationships between analog video timing and the information in the digital timing reference signals for 625/50 (above) and 525/60 (below).

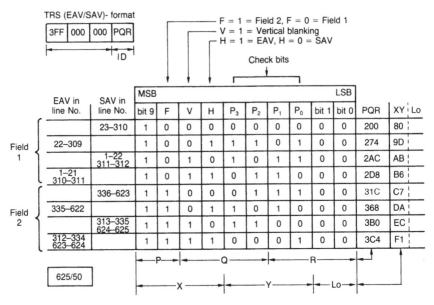

Figure 7.14 Decode table for component TRS.

this capacity is used. In 625 lines, only the active line period may be used on lines 20 to 22 and 333 to 335[5]. Lines 20 and 333 are reserved for equipment self-testing.

In 525 lines there is considerably more freedom and ancillary data may be inserted anywhere where there is no active video, either during horizontal blanking where it is known as HANC, vertical blanking where it is known as VANC, or both[4]. The spare capacity allows many channels of digital audio and considerably simplifies switching.

The all zeros and all ones codes are reserved for synchronizing, and cannot be allowed to appear in ancillary data. In practice only seven bits of the eight-bit word can be used as data; the eighth bit is redundant and gives the byte odd parity. As all ones and all zeros are even parity, the sync pattern cannot then be generated accidentally.

Ancillary data is always prefaced by a different four-symbol TRS which is the inverse of the video TRS in that it starts with all zeros and then has two symbols of all ones followed by the information symbol. See section 8.7 for treatment of embedded audio in SDI.

7.8 The SD parallel interface

Composite digital signals use the same electrical and mechanical interface as used for 4:2:2 component working[4,7]. This means that it is possible erroneously to plug a component signal into a composite machine. Whilst this cannot possibly work, no harm will be done because the signal levels and pinouts are the same.

A 25 pin D-type connector to ISO 2110-1989 is specified. Equipment always has female connectors, cables always have male connectors. Metal or metallized backshells are recommended with screened cables for optimum shielding. Equipment has been seen using ribbon cables and IDC (insulation displacement connectors), but is it not clear whether such cables would meet the newer more stringent EMC (electromagnetic compatibility) regulations. It should be borne in mind that the ninth and eighteenth harmonics of 13.5 MHz are both emergency frequencies for aircraft radio.

Whilst equipment may produce or accept only eight-bit data, cables must contain conductors for all ten bits. Connector latching is by a pair of 4-40 (an American thread) screws, with suitable posts provided on the female connector. It is important that the screws are used as the multicore cable is quite stiff and can eventually unseat the plug if it is not secured. Some early equipment had slidelocks instead of screw pillars, but these proved to be too flimsy. During the changeover from slidelocks to 4-40 screws, some equipment was made with metric screw pillars and these will need to be changed to attach modern cables.

When unscrewing the locking screws from a D-connector it is advisable to check that the lock screw is actually unscrewing from the pillar. It is not unknown for the pillar to rotate instead. If this is not noticed, the pillar fixings may become detached inside the equipment, which will then have to be dismantled.

Each signal in the interface is carried by a balanced pair using ECL (emitter coupled logic) drive levels. The cable has a nominal impedance of 110 ohm and must be correctly terminated. ECL runs from a power supply of nominally -5.2 V and the logic states are -0.8 V for a 'high' and -1.85 V for a 'low'. ECL is primarily a current driven system, and the signal amplitude is quite low compared with other logic families as well as being negative valued.

Figure 7.15 shows the pinouts used. Although it is not obvious from the figure, the numbering of the D-connector is such that signal pairs are on physically opposite pins. Originally most equipment used eight-bit data and ten-bit working was viewed as an option; this was reflected in the wording of the first standards. However, in order to reflect the increasing quantity of ten-bit equipment now in use, the wording of later standards has subtly changed to describe a ten-bit system in which only eight bits may be used.

In the old specification shown in Figure 7.15(a), there are eight signal pairs and two optional pairs, so that extension to a ten-bit word can be

Old 8-bit system	Connector pin number	New 10-bit system
Clock	1	Clock
System ground A	2	System ground A
Data 7 (MSB)	3	Data 9 (MSB)
Data 6	4	Data 8
Data 5	5	Data 7
Data 4	6	Data 6
Data 3	7	Data 5
Data 2	8	Data 4
Data 1	9	Data 3
Data 0	10	Data 2
Data A	11	Data 1
Data B	12	Data 0
Cable shield	13	Cable shield
Clock return	14	Clock return
System ground B	15	System ground B
Data 7 return	16	Data 9 return
Data 6 return	17	Data 8 return
Data 5 return	18	Data 7 return
Data 4 return	19	Data 6 return
Data 3 return	20	Data 5 return
Data 2 return	21	Data 4 return
Data 1 return	22	Data 3 return
Data 0 return	23	Data 2 return
Data A return	24	Data 1 return
Data B return	25	Data 0 return

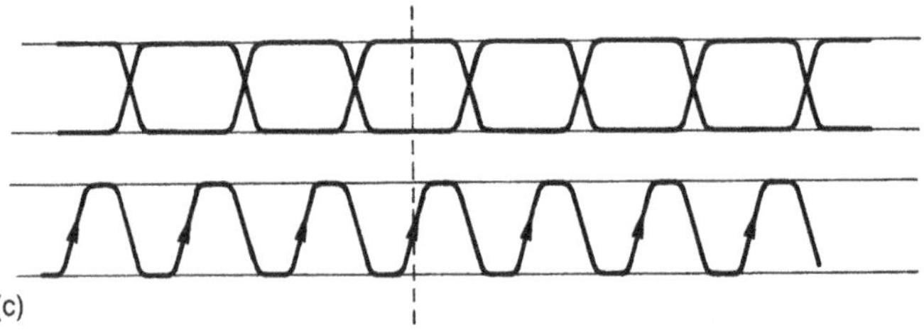

(a)

(b)

(c)

Figure 7.15 The parallel interface was originally specified as in (a) with eight bits expandable to ten. Later documents specify a ten-bit system (b) in which the bottom two bits may be unused. Clock timing at (c) is arranged so that a clock transition occurs between data transitions to allow maximum settling time.

accommodated. The optional signals were used to add bits at the least significant end of the word. Adding bits in this way extends resolution rather than increasing the magnitude. It will be seen from the figure that the optional bits are called Data−1 and Data−2 where the −1 and −2 refer to the powers of two represented, i.e. 2^{-1} and 2^{-2}. The eight-bit word describes 256 levels and ends in a radix point. The extra bits below the radix point represent the half and quarter quantizing intervals. In this way a degree of compatibility exists between ten- and eight-bit systems, as the correct magnitude will always be obtained when changing word length, and all that is lost is a degree of resolution in shortening the word length when the bits below the radix point are lost. The same numbering scheme can be used for both word lengths; the longer word length simply has a radix point and an extra digit in any number base. Converting to the eight-bit equivalent can then be simply a matter of deleting the extra digit and retaining the integer part.

However, the later specification shown in (b) renumbers the bits from 0 to 9 and assumes a system with 1024 levels. Thus all standard levels defined in the old $8 + 2$-bit documents have to be multiplied by four to convert them to the levels in the new ten-bit documents. Thus a level of 16 decimal or 10 hex in the eight-bit system becomes 64 decimal or 40 hex in the ten-bit system. Figure 7.16 shows that $8 + 2$ schemes may use hexadecimal numbering in the XYZ or XYLo format where two hex digits X and Y represent the most significant eight bits and the remaining two bits are represented by the Z or Lo symbol. Ten-bit schemes are numbered with three hex digits PQR where P only has two meaningful bits.

A separate clock signal pair and a number of grounding and shielding pins complete the connection. Figure 7.15 also shows the relationship between the clock and the data. A positive-going clock edge is used to sample the signal lines after the level has settled between transitions. In component, the clock will be line-locked 27 MHz or 36 MHz irrespective of the line standard, whereas in composite digital the clock will be four times the frequency of the PAL or NTSC subcarrier.

The parallel interface is suitable for distances of up to 50 metres (27 MHz) or 40 metres (36 MHz). Beyond this distance equalization is

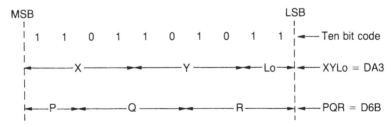

Figure 7.16 As there are two ways to parse ten bits into hexadecimal, there are two numbering schemes in use. In the XYLo system the parsing begins from the MSB down and the Lo parameter is expressed in quarters. In the PQR system the parsing starts from the LSB and so the P parameter has a maximum value of 3.

likely to be necessary and skew or differential delay between signals may become a problem. Equalization of such a large number of signals is not economically viable.

The parallel interface sends active line blocks sandwiched between SAV and EAV TRS codes. The remainder of the blanking periods will contain words whose value alternates between the luma and colour difference blanking codes, unless ancillary data is being sent.

7.9   The HD parallel interface

This obsolescent interface is suitable for RGB or colour difference working. In RGB, each component is carried on three separate sets of conductors whereas in colour difference working the luma is carried on one set and the two colour difference signals are multiplexed into two of the conductor sets, the third being unused, although an auxiliary signal such as a key channel may optionally be sent on the third set. The general concept is identical to the SD parallel interface, with one differential pair of wires per bit and a single differential clock, all at 110 ohm ECL levels. Each pair has its own screen and so three connector pins are required for each bit. Given that three sets of ten-bit data plus a clock are needed, it is clear that the connector will need a massive 93 pins. At a clock rate of 74.25 MHz, this interface is restricted to a length of 20 metres.

As this interface is essentially two, or three for RGB, channels in parallel, each channel has its own synchronizing means. Thus the luma data have their own TRS-ID and the colour difference data also have their own TRS-ID. The TRS codes in each channel should be co-timed.

7.10   The composite digital parallel interface

When composite video is to be digitized, the input will be a single waveform having spectrally interleaved luminance and chroma. Any sampling rate allowing sufficient bandwidth would convey composite video from one point to another. However, if processing in the digital domain is contemplated, there will be less choice.

In the composite digital colour processor it will be necessary to decode the composite signal, which will require some kind of digital filter. Whilst it is possible to construct filters with any desired response, it is a fact that a digital filter whose response is simply related to the sampling rate will be much less complex to implement. This is the reasoning that led to the near universal use of four times subcarrier sampling rate. Figure 7.17 shows the spectra of PAL and NTSC sampled at $4 \times F_{sc}$. It will be evident that there is a considerable space between the edge of the baseband and the lower sideband. This allows the anti-aliasing and reconstruction filters to have a more gradual cut-off, so that ripple in the passband can be reduced. This is particularly important for composite digital recorders, since they are digital devices in an analog environment, and signals may have been converted to and from the digital domain many times in the course of production. A subcarrier multiple sampling clock is easily obtained by gating burst to a phase-locked loop. In NTSC there is no burst swing, whereas at $4F_{sc}$, the burst swing of PAL moves burst crossings by exactly one sample period and so the phase relationship between burst crossings and $4F_s$, clock is unaffected by burst swing.

In NTSC, siting of samples along the line is affected by ScH phase. In PAL, the presence of the 25 Hz component of subcarrier means that samples are not in exactly the same place from one line to the next. The columns lean over slightly such that at the bottom of a field there is a displacement of two samples with respect to the top.

Composite digital samples at four times subcarrier frequency, and so there will be major differences between the PAL and NTSC standards. It is not possible to transmit digitized SECAM. Whilst the component interface transmits only active lines and special sync patterns, the composite interfaces carry the entire composite waveform – syncs, burst and active line. Although ancillary data may be placed in sync tip, the rising and falling sync edges must be present. In the absence of ancillary data, the

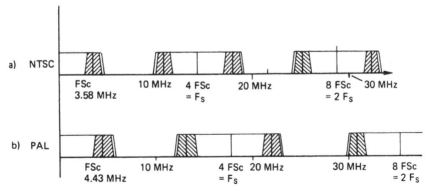

Figure 7.17 The spectra of NTSC at (a) and of PAL at (b) where both are sampled at four times the frequency of their respective subcarriers. This high sampling rate is unnecessary to satisfy sampling theory, and so both are oversampled systems. The advantages are in the large spectral gap between baseband and sideband which allows a more gentle filter slope to be employed, and in the relative ease of colour processing at a sampling rate related to subcarrier.

data on the parallel interface is essentially the continuous stream of samples from a convertor which is digitizing a normal analog composite signal. Virtually all that is necessary to return to the analog domain is to strip out ancillary data and substitute sync tip values prior to driving a DAC and a filter. One of the reasons for this different approach is that the sampling clock in composite video is subcarrier locked. The sample values during sync can change with ScH phase in NTSC and PAL and change with the position in the frame in PAL due to the 25 Hz component. It is simpler to convey sync sample values on the interface than to go to the trouble of recreating them later.

The instantaneous voltage of composite video can go below blanking on dark saturated colours, and above peak white on bright colours. As a result the quantizing ranges need to be stretched in comparison with component in order to accommodate all possible voltage excursions. Sync tip can be accommodated at the low end and peak white is some way below the end of the scale. It is not so easy to determine when overload clipping will take place in composite as the sample sites are locked to subcarrier. The degree of clipping depends on the chroma phase. When samples are taken either side of a chroma peak, clipping will be less likely to occur than when the sample is taken at the peak. Advantage is taken of this phenomenon in PAL as the peak analog voltage of a 100% yellow bar goes outside the quantizing range. The sampling phase is such that samples are sited either side of the chroma peak and remain within the range.

The PAL and NTSC versions of the composite digital interface will be described separately. The electrical interface is the same as for a digital component.

7.10.1 PAL interface

The quantizing range of digital PAL is shown in Figure 7.18[8]. Blanking level is at 256_{10} (64_{10}) and sync tip is the lowest allowable code of 4(1) as 0 is reserved for digital synchronizing. Peak white is 844_{10} (211_{10}).

In PAL, the composite digital interface samples at $4 \times F_{sc}$, with sample phase aligned with burst phase. PAL burst swing results in burst phases of ±135 degrees, and samples are taken at these phases and at ±45 degrees, precisely half-way between the U and V axes. This sampling phase is easy to generate from burst and avoids premature clipping of chroma. It is most important that samples are taken exactly at the points specified, since any residual phase error in the sampling clock will cause the equivalent of a chroma phase error when samples from one source are added to samples from a different source in a switcher. A digital switcher can

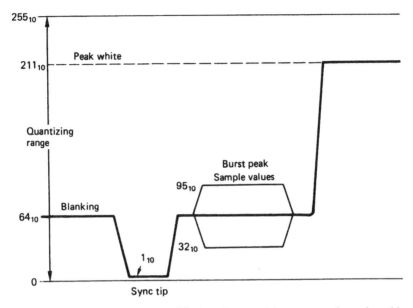

Figure 7.18 The composite PAL signal fits into the quantizing range as shown here. Note that there is sufficient range to allow the instantaneous voltage to exceed that of peak white in the presence of saturated bright colours. Values shown are decimal equivalents in a ten- or eight-bit system. In a ten-bit system the additional two bits increase resolution, not magnitude, so they are below the radix point and the decimal equivalent is unchanged. PAL samples in phase with burst, so the values shown are on the burst peaks and are thus also the values of the envelope.

only add together pairs of samples from different inputs, but if these samples were not taken at the same instants with respect to their subcarriers, the samples represent different vectors and cannot be added.

Figure 7.19 shows how the sampling clock may be derived. The incoming sync is used to derive a burst gate, during which the samples of burst are analysed. If the clock is correctly phased, the sampled burst will give values of 380_{10} (95_{10}), 256_{10} (64_{10}), 128_{10} (32_{10}), 256_{10} (64_{10}) repeated, whereas if a phase error exists, the values at the burst crossings will be above or below 256_{10} (64_{10}). The difference between the sample values and blanking level can be used to drive a DAC that controls the sampling VCO. In this way any phase errors in the ADC are eliminated, because the sampling clock will automatically servo its phase to be identical to digital burst. Burst swing causes the burst peak and burst crossing samples to change places, so a phase comparison is always possible during burst. DC level shifts can be removed by using both positive and negative burst crossings and averaging the results. This also has the effect of reducing the effect of noise.

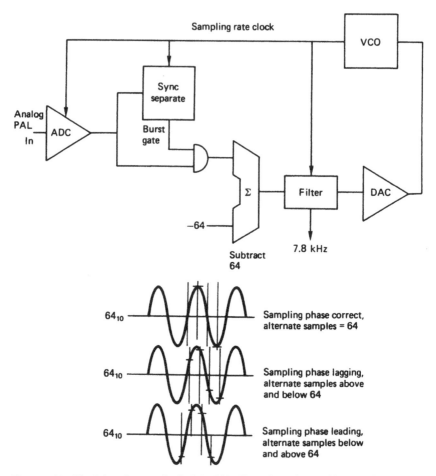

Figure 7.19   Obtaining the sample clock in PAL. The values obtained by sampling burst are analysed. When phase is correct, burst will be sampled at zero crossing and sample value will be 64_{10} or blanking level. If phase is wrong, sample will be above or below blanking. Filter must ignore alternate samples at burst peaks and shift one sample every line to allow for burst swing. It also averages over several burst crossings to reduce jitter. Filter output drives DAC and thus controls sampling clock VCO.

In PAL, the subcarrier frequency contains a 25 Hz offset, and so $4 \times F_{sc}$ will contain a 100 Hz offset. The sampling rate is not h-coherent, and the sampling structure is not quite orthogonal. As subcarrier is given by:

$$F_{sc} = 283.75 \times F_h + F_v/2$$

the sampling rate will be given by:

$$F_s = 1135F_h + 2F_v$$

This results in 709 379 samples per frame, and there will not be a whole number of samples in a line. In practice, 1135 sample periods, numbered 0 to 1134, are defined as one digital line, with an additional 2 sample periods per field which are included by having 1137 samples, numbered 0 to 1136, in lines 313 and 625. Figure 7.20(a) shows the sample numbering scheme for an entire line. Note that the sample numbering begins at 0 at the start of the digital active line so that the horizontal blanking area is near the end of the digital line and the sample numbers will be large. The digital active line is 948 samples long and is longer than the analog active line. This allows the digital active line to move with 25 Hz whilst ensuring the entire analog active line is still conveyed.

Since sampling is not h-coherent, the position of sync pulses will change relative to the sampling points from line to line. The relationship

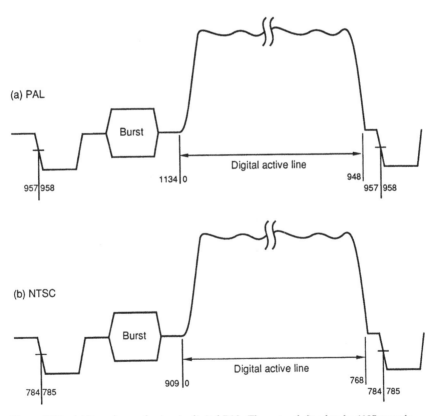

Figure 7.20 (a) Sample numbering in digital PAL. There are defined to be 1135 sample periods per line of which 948 are the digital active line. This is longer than the analog active line. Two lines per frame have two extra samples to compensate for the 25 Hz offset in subcarrier. NTSC is shown at (b). Here there are 910 samples per line of which 768 are the digital active line.

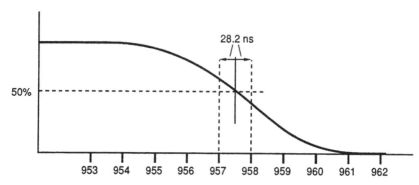

Figure 7.21 As PAL is sampled half-way between the colour axes, sample sites will fall either side of 50% sync at the zero ScH measurement line.

can also be changed by the ScH phase of the analog input. Zero ScH is defined as coincidence between sync and zero degrees of subcarrier phase at line 1 of field 1. Since composite digital samples on burst phase, not on subcarrier phase, the definition of zero ScH will be as shown in Figure 7.21, where it will be seen that two samples occur at exactly equal distances either side of the 50% sync point. If the input is not zero ScH, the samples conveying sync will have different values. Measurement of these values will allow ScH phase to be computed.

7.10.2 NTSC interface

Although they have some similarities, PAL and NTSC are quite different when analysed at the digital sample level. Figure 7.22 shows how the NTSC waveform fits into the quantizing structure[9]. Blanking is at 240_{14} (60_{10}) and peak white is at 800_{10} (200_{10}), so that 1 IRE unit is the equivalent of 1.4Q which could perhaps be called the DIRE. These different values are due to the different sync/vision ratio of NTSC. PAL is 7:3 whereas NTSC is 10:4.

Subcarrier in NTSC has an exact half-line offset, so there will be an integer number of cycles of subcarrier in two lines. F_{sc} is simply $227.5 \times F_h$, and as sampling is at $4 \times F_{sc}$, there will be $227.5 \times 4 = 910$ samples per line period, and the sampling will be orthogonal. Figure 7.20(b) shows that the digital active line consists of 768 samples numbered 0 to 767. Horizontal blanking follows the digital active line in sample numbers 768 to 909.

The sampling phase is chosen to facilitate encoding and decoding in the digital domain. In NTSC there is a phase shift of 123 degrees between subcarrier and the I axis. As burst is an inverted piece of the subcarrier waveform, there is a phase shift of 57 degrees between burst and the I axis. Composite digital NTSC does not sample in phase with burst, but on the I and Q axes at 57, 147, 237 and 327 degrees with respect to burst.

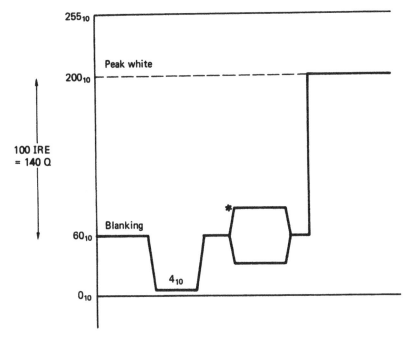

Figure 7.22 The composite NTSC signal fits into the quantizing range as shown here. Note that there is sufficient range to allow the instantaneous voltage to exceed peak white in the presence of saturated, bright colours. Values shown are decimal equivalents in an eight- or ten-bit system. In a ten-bit system the additional two bits increase resolution, not magnitude, so they are below the radix point and the decimal equivalent is unchanged.
* Note that, unlike PAL, NTSC does not sample on burst phase and so values during burst are not shown here. See Figure 7.23 for burst sample details.

Figure 7.23 shows how this approach works in relation to sync and burst. Zero ScH is defined as zero degrees of subcarrier at the 50% point on sync, but the 57 degree sampling phase means that the sync edge is actually sampled 25.6 ns ahead of, and 44.2 ns after, the 50% point. Similarly, when the burst is reached, the phase shift means that burst sample values will be 46_{10}, 83_{10}, 74_{10} and 37_{10} repeating. The phase-locked loop that produces the sampling clock will digitally compare the samples of burst with the values given here. As the burst is not sampled at a zero crossing, the slope will be slightly less. The gain of the phase error detector will also be less, and more prone to burst noise than in the PAL process. The phase error will normally be averaged over several burst samples to overcome this problem.

7.11 Serial digital video interfaces

The serial interfaces described here have a great deal of commonality. Any differences will be noted subsequently. All of them allow up to ten-bit

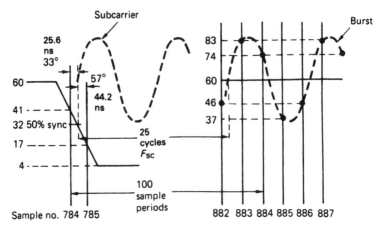

Figure 7.23 NTSC ScH phase. Sampling is not performed in phase with burst as in PAL, but on the *I* and *Q* axes. Since in NTSC there is a phase angle of 57° between burst and *I*, this will also be the phase at which burst samples should be taken. If ScH phase is zero, then phase of subcarrier taken at 50% sync will be zero, and the samples will be taken 33° before and 57° after sync; 25 cycles of subcarrier or 100 samples later, during burst, the sample values will be obtained. Note that in NTSC burst is inverted subcarrier, so sample 785 is positive, but sample 885 is negative.

samples to be communicated serially[10]. If there are only eight bits in the input samples, the missing bits are forced to zero for transmission except for the all-ones condition during TRS which will be forced to ten ones. The interfaces are transparent to ancillary data in the parallel domain, including conveyance of AES/EBU digital audio channels.

Serial transmission uses concepts that were introduced in Chapter 3. At the high bit rates of digital video, the cable is a true transmission line in which a significant number of bits are actually in the cable at any one time, having been sent but not yet received. Under these conditions cable loss is significant. These interfaces operate with cable losses up to 30 dB. The losses increase with frequency and so the bit rate in use and the grade of cable employed both affect the maximum distance the signal will safely travel. Figure 7.24 gives some examples of cable lengths that can be used in SD. In HD there is only one bit rate. Using Belden 1649A or equivalent, a distance of 140 m can be achieved.

Serial transmission uses a waveform that is symmetrical about ground and has an initial amplitude of 800 mV pk–pk across a 75 ohm load. This signal can be fed down 75 ohm coaxial cable having BNC connectors. Serial interfaces are restricted to point-to-point links. Unlike analog video practice, serial digital receivers contain correct termination that is permanently present and passive loop-through is not possible. In permanent installations, no attempt should be made to drive more than one load

System	Clock	Fundamental	Crash knee length	Practical length
NTSC Composite	143 MHz	71.5 MHz	400 m	320 m
PAL Composite	177 MHz	88.5 MHz	360 m	290 m
Component 601	270 MHz	135 MHz	290 m	230 m
Component 16:9	360 MHz	180 MHz	210 m	170 m

CABLE: BICC TM3205, PSF1/2, BELDEN 8281
or any cable with a loss of 8.7 dB/100 m at 100 MHz

Figure 7.24 Suggested maximum cable lengths as a function of cable type and data rate to give a loss of no more than 30 dB. It is unwise to exceed these lengths due to the 'crash knee' characteristic of SDI.

using T-pieces as this will result in signal reflections that seriously compromise the data integrity. On the test bench with very short cables, however, systems with all manner of compromises may still function.

The range of waveforms that can be received without gross distortion is quite small and raw data produce waveforms outside this range. The solution is the use of scrambling, or pseudo-random coding. The serial interfaces use convolutional scrambling as was described in Chapter 3. This is simpler to implement in a cable installation because no separate synchronizing of the randomizing is needed. The scrambling process at the transmitter spreads the signal spectrum and makes that spectrum reasonably constant and independent of the picture content. It is possible to assess the degree of equalization necessary by comparing the energy in a low-frequency band with that in higher frequencies. The greater the disparity, the more equalization is needed. Thus fully automatic cable equalization at the receiver is easily achieved.

The essential parts of a serial link are shown in Figure 7.25. Parallel data having a word length of up to ten bits forms the input. These are fed to a ten-bit shift register which is clocked at ten times the input word rate: 1.485 GHz, 360 MHz, 270 MHz or $40 \times F_{sc}$. The serial data emerge from the shift register LSB first and are then passed through the scrambler, in which a given bit is converted to the exclusive-OR of itself and two bits that are five and nine clocks ahead. This is followed by another stage, which converts channel ones into transitions. The transition encoder ensures that the signal is polarity independent. The resulting logic level signal is converted to a 75 ohm source impedance signal at the cable driver.

The receiver must regenerate a bit clock at 1.485 MHz, 360 MHz, 270 MHz or $40 \times F_{sc}$ from the input signal, and this clock drives the input sampler and

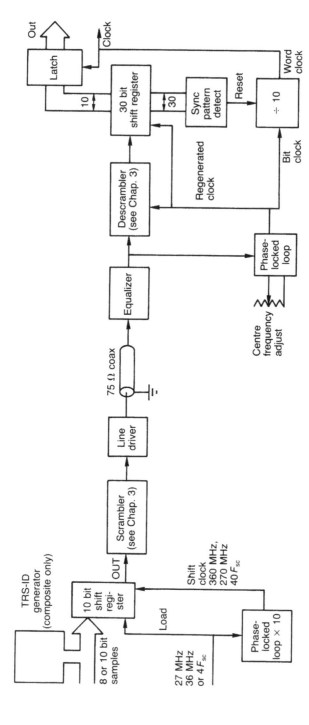

Figure 7.25 Major components of an SDI link. See text for details.

slicer which converts the cable waveform back to serial binary. The local bit clock also drives a circuit that simply reverses the scrambling at the transmitter. The first stage returns transitions to ones. The second stage is a mirror image of the encoder and reverses the exclusive-OR calculation to output the original data. Such descrambling results in error extension, but this is not a practical problem since link error rates are practically zero.

As transmission is serial, it is necessary to obtain word synchronization, so that correct deserialization can take place. The TRS patterns are used for this purpose. The all-ones and all-zeros bit patterns form a unique 30-bit sequence which is detected in the receiver's shift register. The transition from all ones to all zeros is on a word boundary and from that point on the deserializer simply divides by ten to find the word boundaries in the transmission.

7.11.1 *Standard definition serial digital interface (SDI)*

This interface supports 525/59.94 2:1 and 625/50 2:1 scanning standards in component and composite. The component interfaces use a common bit rate of 270 MHz for 4:3 pictures with an option of 360 MHz for 16:9. In component, the TRS codes are already present in the parallel domain and SDI does no more than serialize the parallel signal protocol unchanged.

Composite digital samples at four times the subcarrier frequency and so the bit rate is different between the PAL and NTSC variants. The composite parallel interface signal is not a multiplex and also carries digitized analog syncs. Consequently there is no need for TRS codes. For serial transmission it is necessary to insert TRS at the serializer and subsequently to strip it out at the serial-to-parallel convertor. The TRS-ID is inserted during blanking, and the serial receiver can detect the patterns it contains. Composite TRS-ID is different to the one used in component signals and consists of five words inserted just after the leading edge of analog video sync. Figure 7.26(a) shows the location of TRS-ID at samples 967–971 in PAL and (b) shows the location at samples 790–794 in NTSC.

Out of the five words in TRS-ID, the first four are for synchronizing, and consist of a single word of all ones, followed by three words of all zeros. Note that the composite TRS contains an extra word of zeros compared with the component TRS and this could be used for signal identification in multi-standard devices. The fifth word is for identification, and carries the line and field numbering information shown in Figure 7.27. The field numbering is colour-framing information useful for editing. In PAL the field numbering will go from zero to seven, whereas in NTSC it will only reach three.

(a) PAL

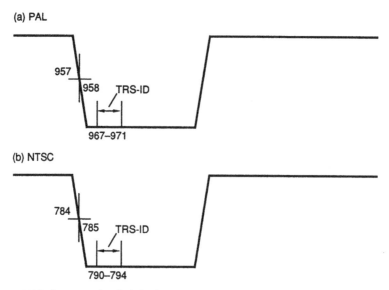

(b) NTSC

Figure 7.26 In composite digital it is necessary to insert a sync pattern during analog sync tip to ensure correct deserialization. The location of TRS-ID is shown at (a) for PAL and at (b) for NTSC.

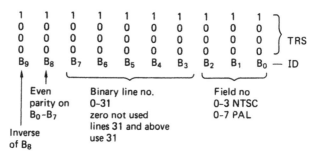

Figure 7.27 The contents of the TRS-ID pattern which is added to the transmission during the horizontal sync pulse just after the leading edge. The field number conveys the composite colour framing field count, and the line number carries a restricted line count intended to give vertical positioning information during the vertical interval. This count saturates at 31 for lines of that number and above.

On detection of the synchronizing symbols, a divide-by-ten circuit is reset, and the output of this will clock words out of the shift register at the correct times. This circuit will also provide the output word clock.

7.11.2 SDTI

SDI is closely specified and is only suitable for transmitting 2:1 interlaced 4:2:2 digital video in 525/60 or 625/50 systems. Since the development of

SDI, it has become possible economically to compress digital video and the SDI standard cannot handle this. SDTI (serial data transport interface) is designed to overcome that problem by converting SDI into an interface that can carry a variety of data types whilst retaining compatibility with existing SDI router infrastructures.

SDTI sources produce a signal which is electrically identical to an SDI signal and which has the same timing structure. However, the digital active line of SDI becomes a data packet or item in SDTI. Figure 7.28 shows how SDTI fits into the existing SDI timing. Between EAV and SAV (horizontal blanking in SDI) an ancillary data block is incorporated. The structure of this meets the SDI standard, and the data within describes the contents of the following digital active line.

The data capacity of SDTI is about 200 Mbits/s because some of the 270 Mbits/s are lost due to the retention of the SDI timing structure. Each digital active line finishes with a CRCC (cyclic redundancy check character) to check for correct transmission.

SDTI raises a number of opportunities, including the transmission of compressed data at faster than real time. If a video signal is compressed at 4:1, then one quarter as much data would result. If sent in real time the bandwidth required would be one quarter of that needed by uncompressed video. However, if the same bandwidth is available, the compressed data

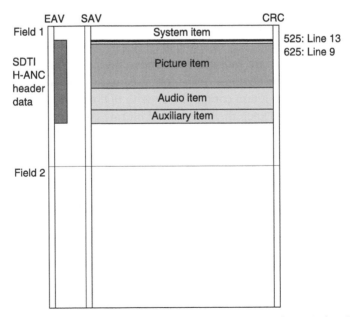

Figure 7.28 SDTI is a variation of SDI which allows transmission of generic data. This can include compressed video and non-real-time transfer.

could be sent in one quarter of the usual time. This is particularly advantageous for data transfer between compressed camcorders and non-linear editing workstations. Alternatively, four different 50 Mbit/s signals could be conveyed simultaneously.

Thus an SDTI transmitter takes the form of a multiplexer which assembles packets for transmission from input buffers. The transmitted data can be encoded according to MPEG, MotionJPEG, Digital Betacam or DVC formats and all that is necessary is that compatible devices exist at each end of the interface. In this case the data are transferred with bit accuracy and so there is no generation loss associated with the transfer. If the source and destination are different, that is, having different formats or, in MPEG, different group structures, then a conversion process with attendant generation loss would be needed.

7.11.3 ASI

The asynchronous serial interface is designed to allow MPEG transport streams to be transmitted over standard SDI cabling and routers. ASI offers higher performance than SDTI because it does not adhere to the SDI timing structure. Transport stream data do not have the same statistics as PCM video and so the scrambling technique of SDI cannot be used. Instead ASI uses an 8/10 group code (see section 3.8) to eliminate DC components and ensure adequate clock content.

SDI equipment is designed to run at a closely defined bit rate of 270 Mbits/s and has phase-locked loops in receiving and repeating devices which are intended to remove jitter. These will lose lock if the channel bit rate changes. Transport streams are fundamentally variable in bit rate and to retain compatibility with SDI routing equipment ASI uses stuffing bits to keep the transmitted bit rate constant.

The use of an 8/10 code means that although the channel bit rate is 270 Mbits/s, the data bit rate is only 80% of that, that is, 216 Mbits/s. A small amount of this is lost to overheads.

7.11.4 High definition serial digital interface (HD-SDI)

The SD serial interface runs at a variety of bit rates according to the television standard being sent. In contrast the HD serial interface[11] runs at only one bit rate, 1.485 Gbits/s, although it is possible to reduce this by 0.1% so that it can lock to traditional 59.94 Hz equipment. At this high bit rate, variable speed causes too many difficulties and it is easier to accommodate a reduced data rate by sending more blanking or ancillary data so that the transmitted bit rate stays the same. A receiver can work out which format is being sent by counting the number of blanking periods between the active lines.

Apart from the bit rate, the HD serial interface has as much in common with the SDI standard as possible. Although the impedance, signal level and channel coding are the same, the HD serial interface has a number of detail differences in the protocol.

The parallel HD interface above has two channels, one for luma and one for multiplexed colour difference data. Each of these has a symbol rate of 74.25 MHz and has its own TRS-ID structure. Essentially the HD serial interface is transparent to this data as it simply multiplexes between the two channels at symbol rate. As far as the active line is concerned, the result is the same as for SD: a sequence of C_b, Y, C_r, Y, etc. However, in HD the TRS-IDs of the two channels are also multiplexed. A further difference is that the HD interface has a line number and a CRC for each active line inserted immediately after EAV. Figure 7.29(a) shows the EAV and SAV structure of each channel, with the line count and CRC, whereas (b) shows the resultant multiplex.

7.12 Digital video interfacing chipsets

Implementation of digital video systems is much easier now that specialized chips are available. The introduction of HD-SDI has required significant increase in chip performance to support the additional bit rate. HD chips are thus more expensive than their SD equivalent. One useful move is that HD and SD chips are being made with the same pinouts. Thus a

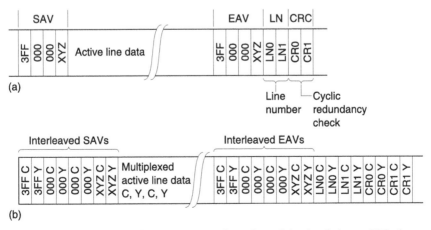

Figure 7.29   The HD parallel data are in two channels, each having their own TRS, shown at (a). The EAV is extended by line number and CRC. (b) When the two channels are multiplexed, the TRS codes are interleaved as shown.

single circuit board can be made into an SD or an HD device just by installing chips of the appropriate speed.

Figure 7.30 shows a hypothetical 4:2:2 component system starting with analog signals and ending with the same to illustrate the processes which are necessary. The syncs on Y are separated and multiplied in a phase-locked loop to produce a 27 MHz master clock. This is divided by 2 and by 4 to produce the sampling clocks for the convertors. This results in three data streams, which can be multiplexed to form a parallel interface signal using a parallel encoder chip such as the Sony CXD8068G. This parallel signal may be output using a set of ECL line drivers. If it is required to convert the parallel signal to SDI, a serial encoder will be required. The Sony SBX1610A and the Gennum GS9002 contain all parallel-to-serial functions, but output logic level signals which require a CXA 1389AQ or a GS9007 cable driver to produce the 1.6 volt pk–pk SDI signal which will fall to the standard 0.8 volts after passing through the source terminating resistors.

At the receiving end of the cable the signal requires equalization, clock regeneration and deserializing. The Sony SBX1602A provides all of these functions in one device whereas the Gennum solution is to combine equalization and reclocking in the GS9005 and to perform deserialization in the GS9000. In both cases the output is parallel single-ended data which can be returned to the parallel interface specification using ECL

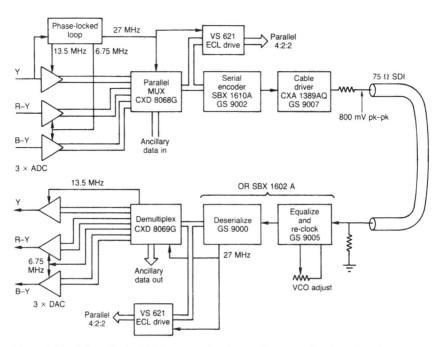

Figure 7.30 A hypothetical 4:2:2 system showing applications of various digital interfacing chips.

drivers. Alternatively the parallel data may be sent directly to a parallel interface decoder such as the Sony CXD8069G which demultiplexes the 27 MHz data to provide separate outputs for driving three DACs.

Figure 7.31 shows a block diagram of the CXD8068G parallel interface encoder. This accepts the parallel input from three component ADCs and multiplexes them to the 27 MHz parallel standard. The rounding process allows ten-bit inputs to be rounded to shorter word lengths. The limiter prevents out-of-range analog signals from producing all-ones or all-zeros codes which are reserved for synchronizing. In addition to a 27 MHz clock derived from horizontal sync, the chip requires horizontal and frame drives to operate the timing counters which address the TRS generator. The final multiplexer selects TRS patterns, video data or ancillary data for the ten-bit parallel output.

Figure 7.32(a) shows the SBX1601A serial encoder and (b) shows the GS9002 serial encoder. Of necessity these chips contain virtually identical processing. Parallel input data are clocked into the input latch by the parallel word clock which is multiplied in frequency by a factor of ten in a phase-locked loop to provide a serial bit clock. There is provision for selecting several centre frequencies for composite or component applications. The data latch output is examined by logic that detects input sync patterns and extends eight-bit sync values to ten bits. The parallel data are then serialized in a shift register prior to passing through the scrambler and the transition generator.

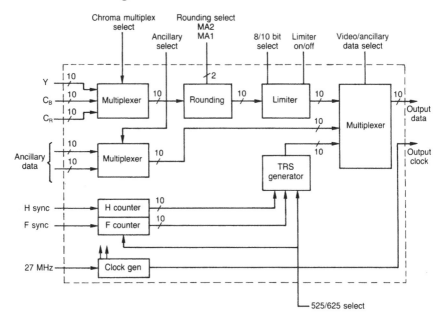

Figure 7.31 The three data streams from component ADCs can be multiplexed into the parallel interface format with the encoder chip shown here.

Figure 7.33 shows an SDI cable driver chip. The device shown has quadruple outputs and is useful in applications such as distribution amplifiers. Note that each differential amplifier produces a pair of separate SDI outputs. The fact that these are mutually inverted is irrelevant as the SDI signal is not polarity conscious. Note the resistor networks that provide correct cable source termination.

Figure 7.34(a) shows the Gennum GS9005 reclocking receiver. This contains an automatic cable equalizer and a phase-locked loop clock recovery circuit that drives a slicer/sampler to recover the channel waveform to a logic level signal for subsequent descrambling in a separate device. The equalizer operates by estimating the cable length from the input amplitude and driving a voltage-controlled filter from the signal strength. A buffered eye pattern test point is provided. The equalizer output is DC restored prior to slicing to ensure that the slicing takes place around the waveform centre line. The slicer output will contain timing jitter and so a phase-locked loop is used having a loop filter to reject the jitter. The jitter-free clock is used to drive the data latch which samples the slicer output between transitions. The VCO centre frequency can be selected from four values and provision is made for an adjusting potentiometer for each frequency.

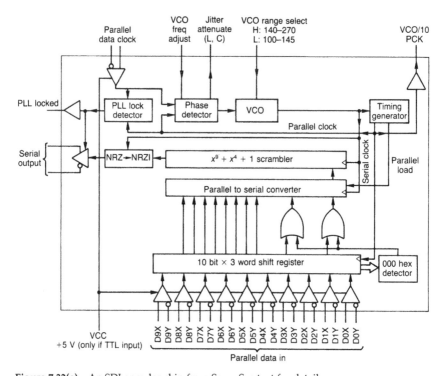

Figure 7.32(a)   An SDI encoder chip from Sony. See text for details.

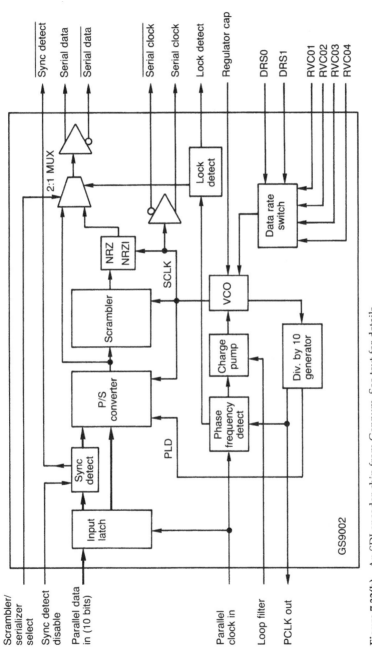

Figure 7.32(b) An SDI encoder chip from Gennum. See text for details.

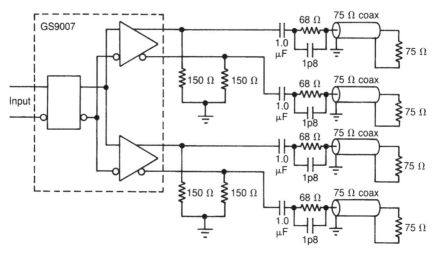

Figure 7.33   SDI cable driver chip provides correct 0.8 V signal level after source termination resistors.

Figure 7.34(b) shows the GS9000 serial decoder which complements the GS9005. This contains a descrambler and a serial-to-parallel convertor synchronized by the detection of TRS in the shift register. The chip also contains an automatic standard detector that outputs a two-bit standard code for external indication and to select the centre frequency of the GS9005. The single-ended parallel output can be converted to the differential parallel output standard using a multiple ECL driver such as a VS621.

Figure 7.35 shows the Sony SBX1602A, which contains all of the serial receiving functions in one device. Its operation should be self-evident from the description of the Gennum devices above.

Parallel data can be demultiplexed for conversion to analog by the CXD8069G device shown in Figure 7.36 that also extracts ancillary data. The TRS detector identifies sync patterns and uses them to direct the ID word to the Hamming code error-correction stage. This outputs corrected timing signals that are decoded to produce analog video timing drives. A FIFO (First in First out) buffer acts as a small timebase corrector to allow the DACs to be driven with a stable clock. Ten-bit video data may be rounded to shorter word lengths if required, prior to demultiplexing into separate component outputs.

As the HD protocol is based heavily on the SD protocol, HD chipsets differ primarily in the bit rate they can handle. Detail differences include the generation of the line count parameter and CRC following SAV and the need for a different sync recognition system owing to the interleaving of two TRS codes in the serial bitstream. Figure 7.37 shows a typical HD serial system.

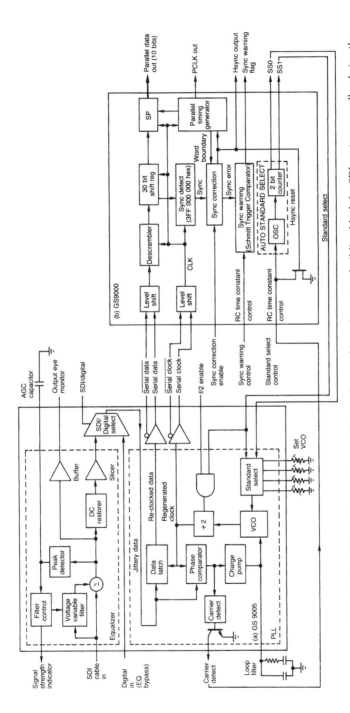

Figure 7.34 (a) Reclocking SDI receiver contains a cable equalizer and is an important building block for SDI routers as well as being the fast stage of an SDI decoder. Decoder is shown in (b). Note auto standard sensing outputs which can select VCO frequency in the reclocker.

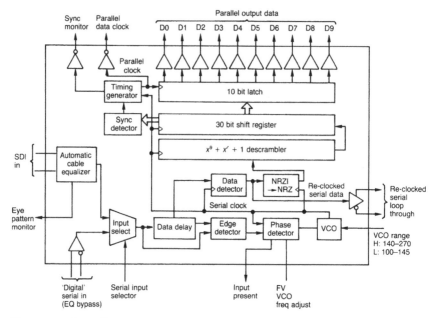

Figure 7.35 Sony SDI receiver chip for comparison with Figure 7.34.

7.13 Embedded audio in SDI

In component SDI, there is provision for ancillary data packets to be sent during blanking[10,12]. The high clock rate of component means that there is capacity for up to 16 audio channels sent in four groups. Composite SDI has to convey the digitized analog sync edges and bursts and only sync tip is available for ancillary data. As a result of this and the lower clock rate, composite has much less capacity for ancillary data than component although it is still possible to transmit one audio data packet carrying four audio channels in one group. Figure 7.38(a) shows where the ancillary data may be located for PAL and (b) shows the locations for NTSC.

As was shown in Chapter 4, the data content of the AES/EBU digital audio subframe consists of validity (V), user (U) and channel (C) status bits, a 20-bit sample and four auxiliary bits which optionally may be appended to the main sample to produce a 24-bit sample. The AES recommends sampling rates of 48, 44.1 and 32 kHz, but the interface permits variable sampling rates. SDI has various levels of support for the wide range of audio possibilities and these levels are defined in Figure 7.39. The default or minimum level is Level A which operates only with a video-synchronous 48 kHz sampling rate and transmits V, U, C and the main 20-bit sample only. As Level A is a default it need not be signalled to a receiver as the presence of IDs in the ancillary data is enough to

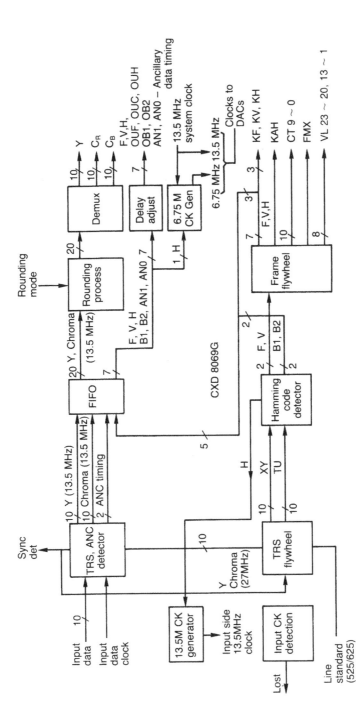

Figure 7.36 This device demultiplexes component data to drive separate DACs for each component as well as stripping out ancillary data.

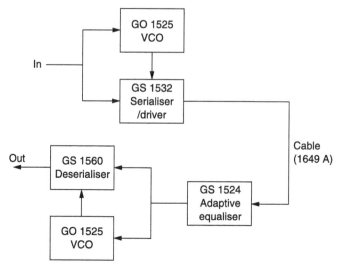

Figure 7.37 Components of a typical HD-SDI system.

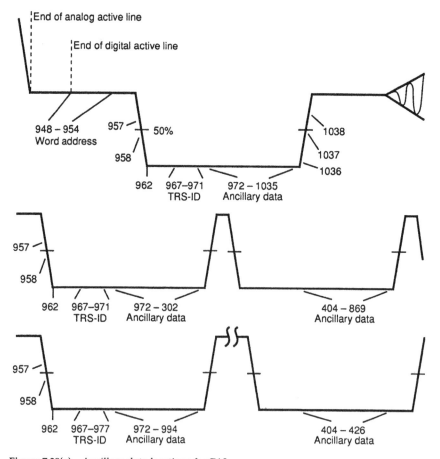

Figure 7.38(a) Ancillary data locations for PAL.

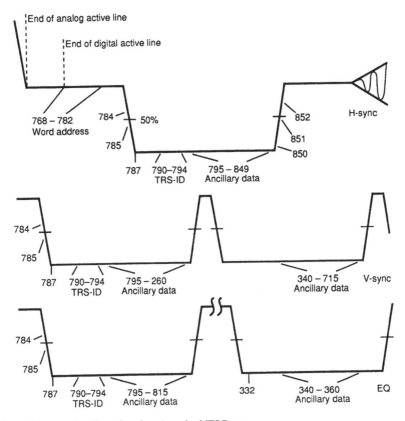

Figure 7.38(b) Ancillary data locations for NTSC.

	A (Default)	Synchronous 48 kHz, 20 bit audio, 48 sample buffer
	B	Synchronous 48 kHz for composite video. 64 sample buffer to receive
		20 bits from 24 bit data
	C	Synchronous 48 kHz 24 bit with extended packets
	D	Asynchronous audio
	E	44.1 kHz audio
	F	32 kHz audio
	G	32–48 kHz variable sampling rate
	H	Audio frame sequence
	I	Time delay tracking
	J	Non-coincident channel status Z bits in a pair

Needs audio control packet

Figure 7.39 The different levels of implementation of embedded audio. Level A is default.

ensure correct decoding. However, all other levels require an audio control packet to be transmitted to teach the receiver how to handle the embedded audio data. The audio control packet is transmitted once per field in the second horizontal ancillary space after the video switching point before any associated audio sample data. One audio control packet is required per group of audio channels.

If it is required to send 24-bit samples, the additional four bits of each sample are placed in extended data packets that must directly follow the associated group of audio samples in the same ancillary data space.

There are thus three kinds of packet used in embedded audio: the audio data packet which carries up to four channels of digital audio, the extended data packet and the audio control packet.

In component systems, ancillary data begins with a reversed TRS or sync pattern. Normal video receivers will not detect this pattern and so ancillary data cannot be mistaken for video samples. The ancillary data TRS consists of all zeros followed by all ones twice. There is no separate TRS for ancillary data in composite. Immediately following the usual TRS, there will be an ancillary data flag whose value must be $3FC_{16}$. Following the ancillary TRS or data flag is a data ID word containing one of a number of standardized codes which tell the receiver how to interpret the ancillary packet. Figure 7.40 shows a list of ID codes for various types of packets. Next come the data block number and the data block count parameters. The data block number increments by 1 on each instance of a block with a given ID number. On reaching 255 it overflows and recommences counting. Next, a data count parameter specifies how many symbols of data are being sent in this block. Typical values for the data count are 36_{10} for a small packet and 48_{10} for a large packet. These parameters help an audio extractor to assemble contiguous data relating to a given set of audio channels.

Figure 7.41 shows the structure of the audio data packing. In order to prevent accidental generation of reserved synchronizing patterns, bit 9 is the inverse of bit 8 so the effective system word length is nine bits. Three nine-bit symbols are used to convey all of the AES/EBU subframe data except for the four auxiliary bits. Since four audio channels can be conveyed, there are

	Group 1	Group 2	Group 3	Group 4
Audio data	2FF	1FD	1FB	2F9
Audio CTL	1EF	2EE	2ED	1EC
Ext. data	1FE	2FC	2FA	1F8

Figure 7.40 The different packet types have different ID codes as shown here.

Address Bit	x3	x3 + 1	x3 + 2
B9	$\overline{B8}$	$\overline{B8}$	$\overline{B8}$
B8	A (2^5)	A (2^{14})	P
B7	A (2^4)	A (2^{13})	C
B6	A (2^3)	A (2^{12})	U
B5	A (2^2)	A (2^{11})	V
B4	A (2^1)	A (2^{10})	A MSB (2^{19})
B3	A LSB (2^0)	A (2^9)	A (2^{18})
B2	CH (MSB)	A (2^8)	A (2^{17})
B1	CH (LSB)	A (2^7)	A (2^{16})
B0	Z	A (2^6)	A (2^{15})

Figure 7.41 AES/EBU data for one audio sample is sent as three nine-bit symbols. A = audio sample. Bit Z = AES/EBU channel status block start bit.

two 'Ch' or channel number bits specifying the audio channel number to which the subframe belongs. A further bit, Z, specifies the beginning of the 192-sample channel status message. V, U and C have the same significance as in the normal AES/EBU standard, but the P bit reflects parity on the three nine-bit symbols rather than the AES/EBU definition. The three-word sets representing an audio sample will then be repeated for the remaining three channels in the packet but with different combinations of the Ch bits.

One audio sample in each of the four channels of a group requires 12 video sample periods and so packets will contain multiples of 12 samples. At the end of each packet a checksum is calculated on the entire packet contents.

If 24-bit samples are required, extended data packets must be employed in which the additional four bits of each audio sample in an AES/EBU frame are assembled in pairs according to Figure 7.42. Thus for every 12 symbols conveying the four 20-bit audio samples of one group in an audio data packet two extra symbols will be required in an extended data packet.

The audio control packet structure is shown in Figure 7.43. Following the usual header are symbols representing the audio frame number, the sampling rate, the active channels, the processing delay and some reserved symbols. The sampling rate parameter allows the two AES/EBU channel pairs in a group to have different sampling rates if required. The active channel parameter simply describes which channels in a group carry meaningful audio data. The processing delay parameter denotes the delay the audio has experienced measured in audio sample periods. The parameter is a 26-bit two's complement number requiring three symbols for each channel. Since the four audio channels in a group are generally channel pairs, only two delay parameters are needed. However, if four independent channels are used, one parameter each will be required. The e bit denotes whether four individual channels or two pairs are being transmitted.

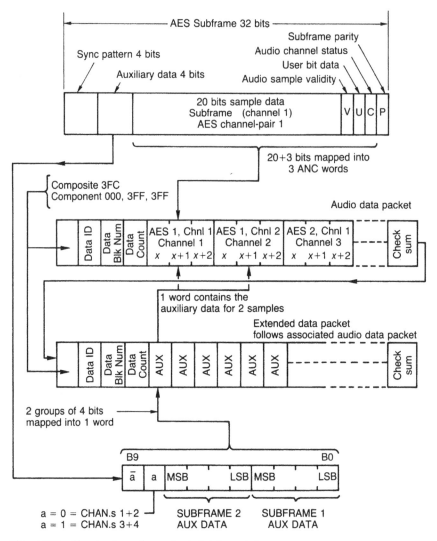

Figure 7.42 The structure of an extended data packet.

The frame number parameter comes about in 525 line systems because the frame rate is 29.97 Hz not 60 Hz. The resultant frame period does not contain a whole number of audio samples. An integer ratio is only obtained over the multiple frame sequence shown in Figure 7.44. The frame number conveys the position in the frame sequence. At 48 kHz odd frames hold 1602 samples and even frames hold 1601 samples in a five-frame sequence. At 44.1 and 32 kHz the relationship is not so simple and to obtain the correct number of samples in the sequence certain frames (exceptions) have the number of samples altered. At 44.1 kHz the frame sequence is 100 frames long whereas at 32 kHz it is 15 frames long.

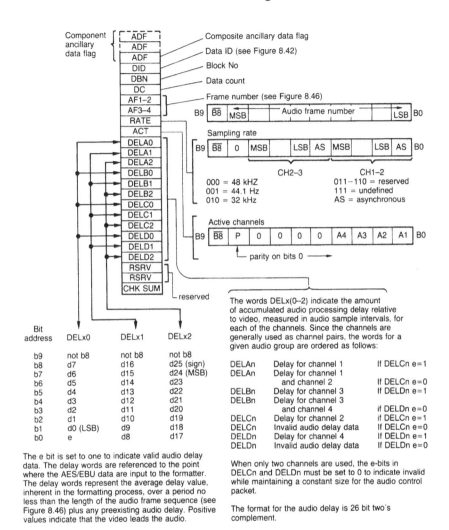

Figure 7.43 The structure of an audio control packet.

As the two channel pairs in a group can have different sampling rates, two frame parameters are required per group. In 50 Hz systems all three sampling rates allow an integer number of samples per frame and so the frame number is irrelevant.

As the ancillary data transfer is in bursts, it is necessary to provide a little RAM buffering at both ends of the link to allow real-time audio samples to be time compressed up to the video bit rate at the input and expanded back again at the receiver. Figure 7.45 shows a typical audio insertion unit in which the FIFO buffers can be seen. In such a system all that matters is that the average audio data rate is correct. Instantaneously there can be timing errors within the range of the buffers. Audio data cannot be embedded at the video switch point or in the areas reserved for

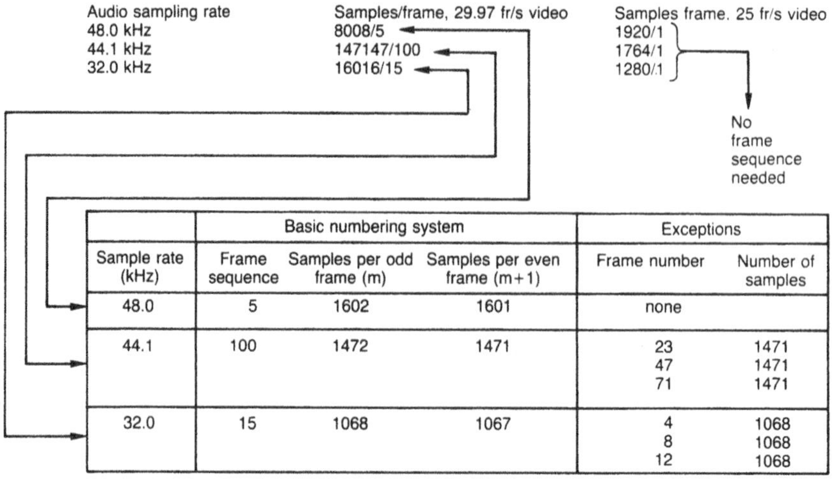
Audio sampling rate
48.0 kHz
44.1 kHz
32.0 kHz
Samples/frame, 29.97 fr/s video
8008/5
147147/100
16016/15
Samples frame. 25 fr/s video
1920/1
1764/1
1280/1
No frame sequence needed

Sample rate (kHz)	Basic numbering system: Frame sequence	Samples per odd frame (m)	Samples per even frame (m+1)	Exceptions: Frame number	Number of samples
48.0	5	1602	1601	none	
44.1	100	1472	1471	23 47 71	1471 1471 1471
32.0	15	1068	1067	4 8 12	1068 1068 1068

Figure 7.44 The origin of the frame sequences in 525 line systems.

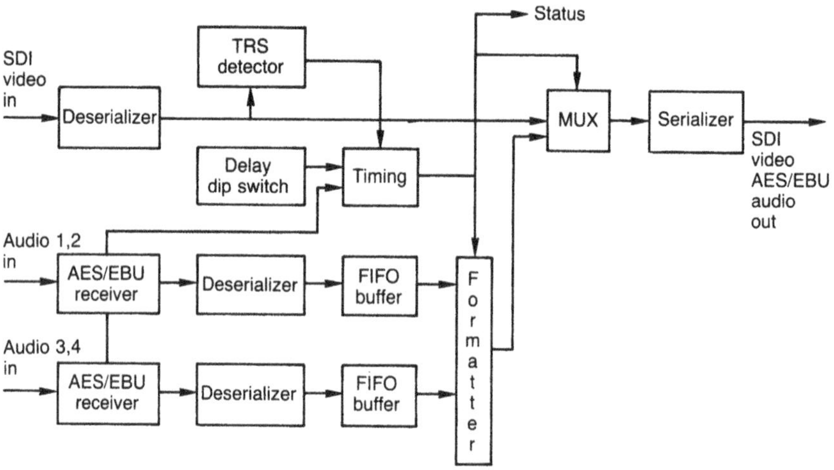
Status
SDI video in
TRS detector
Deserializer
MUX
Serializer
SDI video AES/EBU audio out
Delay dip switch
Timing
Audio 1,2 in
AES/EBU receiver
Deserializer
FIFO buffer
Formatter
Audio 3,4 in
AES/EBU receiver
Deserializer
FIFO buffer

Figure 7.45 A typical audio insertion unit. See text for details.

EDH packets, but provided that data are evenly spread throughout the frame 20-bit audio can be embedded and retrieved with about 48 audio samples of buffering. If the additional four bits per sample are sent this requirement rises to 64 audio samples. The buffering stages cause the audio to be delayed with respect to the video by a few milliseconds at each insertion. Whilst this is not serious, Level I allows a delay-tracking mode which allows the embedding logic to transmit the encoding delay so a subsequent receiver can compute the overall delay. If the range of the buffering is exceeded for any reason, such as a non-synchronous audio sampling rate fed to a Level A encoder, audio samples are periodically skipped or repeated in order to bring the delay under control.

It is permitted for receivers that can only handle 20-bit audio to discard the four-bit sample extension data. However, the presence of the extension data requires more buffering in the receiver. A device having a buffer of only 48 samples for Level A working could experience an overflow due to the presence of the extension data.

In 48 kHz working, the average number of audio samples per channel is just over three per video line. In order to maintain the correct average audio sampling rate, the number of samples sent per line is variable and not specified in the standard. In practice a transmitter generally switches between packets containing three samples and packets containing four samples per channel per line as required to keep the buffers from overflowing. At lower sampling rates either smaller packets can be sent or packets can be omitted from certain lines.

As a result of the switching, ancillary data packets in component video occur mostly in two sizes. The larger packet is 55 words in length of which 48 words are data. The smaller packet contains 43 words of which 36 are data. There is space for two large packets or three small packets in the horizontal blanking between EAV and SAV.

A typical embedded audio extractor is shown in Figure 7.46. The extractor recognizes the ancillary data TRS or flag and then decodes the ID to determine the content of the packet. The group and channel addresses are then used to direct extracted symbols to the appropriate

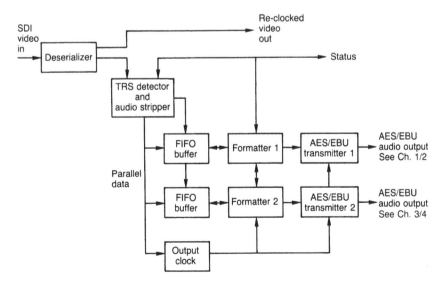

Figure 7.46 A typical audio extractor. Note the FIFOs for timebase expansion of the audio samples.

audio channel. A FIFO memory is used to timebase expand the symbols to the correct audio sampling rate.

7.14 EDH – error detection and handling

Surprisingly, the original SD-SDI standard had no provisions for data integrity checking. EDH is an option for SD-SDI which rectifies the omission[13,14]. Figure 7.47 shows an EDH equipped SDI (serial digital interface) transmission system. At the first transmitter, the data from one field is transmitted and simultaneously fed to a cyclic redundancy check (CRC) generator. The CRC calculation is a mathematical division by a polynomial and the result is the remainder. The remainder is transmitted in a special ancillary data packet sent early during the vertical interval, before any switching takes place in a router[14]. The first receiver has an identical CRC generator that performs a calculation on the received field. The ancillary data extractor identifies the EDH packet and demultiplexes it from the main data stream. The remainder from the ancillary packet is then compared with the locally calculated remainder. If the transmission is error free, the two values will be identical. In this case no further action results. However, if as little as one bit is in error in the data, the remainders will not match. The remainder is a 16-bit word and guarantees to detect up to 16 bits in error anywhere in the field. Greater numbers of errors are not guaranteed to be detected, but this is of little consequence as enough fields in error will be detected to indicate that there is a problem.

Should a CRC mismatch indicate an error in this way, two things happen. First, an optically isolated output connector on the receiving equipment will present a low impedance for a period of 1 to 2 milliseconds. This will result in a pulse in an externally powered circuit to indicate that a field contained an error. An external error-monitoring system wired to this connector can note the occurrence in a log or sound an alarm or whatever it is programmed to do. As the data have been incorrectly received, the fact must also be conveyed to subsequent equipment. It is not permissible to pass on a mismatched remainder. The centre unit in Figure 7.47 must pass on the data as received, complete with errors, but it must calculate a new CRC that matches the erroneous data. When received by the third unit in Figure 7.47, there will then only be a CRC mismatch if the transmission between the second and third devices is in error. This is correct as the job of the CRC is only to locate faulty hardware and clearly if the second link is not faulty the CRC comparison should not fail. However, the third device still needs to know that there is a problem with the data, and this is the job of the error flags that also reside in the EDH packet. One of these flags is called edh (error detected here) and will be asserted by the centre device in Figure 7.47.

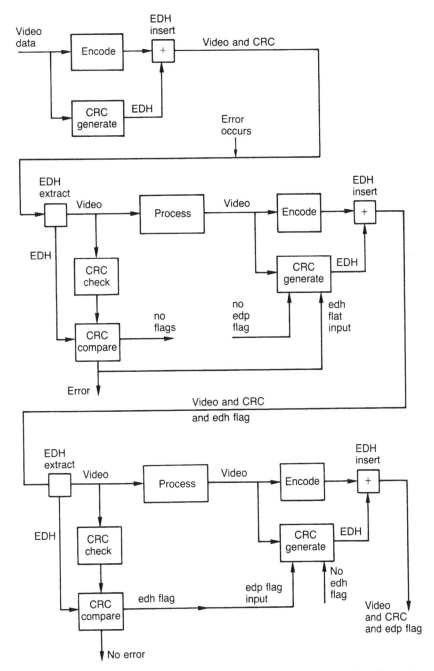

Figure 7.47 A typical EDH system illustrating the way errors are detected and flagged. See text for details.

The last device in Figure 7.47 will receive edh and transmit eda (error detected already). There are also flags to handle hardware failures (e.g. over-temperature or diagnostic failure). The idh (internal error detected here) and ida (internal error detected already) handle this function. Locally detected hardware errors drive the error output socket to a low impedance state constantly to distinguish from the pulsing of a CRC mismatch.

A slight extra complexity is that error checking can be performed in two separate ways. One CRC is calculated for the active picture only, and another is calculated for the full field. Both are included in the EDH packet shown in Figure 7.48. The advantage of this arrangement is that whilst regular programme material is being passed in active picture, test patterns can be sent in vertical blanking which can be monitored separately. Thus

Data item	b9 msb	b8	b7	b6	b5	b4	b3	b2	b1	b0 lsb
Ancillary data header, word 1 – component	0	0	0	0	0	0	0	0	0	0
Ancillary data header, word 2 – component	1	1	1	1	1	1	1	1	1	1
Ancillary data header, word 3 – component	1	1	1	1	1	1	1	1	1	1
Auxiliary data flag – composite	1	1	1	1	1	1	1	1	0	0
Data ID (1F4)	0	1	1	1	1	1	0	1	0	0
Block number	1	0	0	0	0	0	0	0	0	0
Data count	0	1	0	0	0	1	0	0	0	0
Active picture data word 0 crc<5:0>	$\bar{P}$	P	C_5	C_4	C_3	C_2	C_1	C_0	0	0
Active picture data word 1 crc<11:6>	$\bar{P}$	P	C_{11}	C_{10}	C_9	C_8	C_7	C_6	0	0
Active picture data word 2 crc<15:12>	$\bar{P}$	P	V	0	C_{15}	C_{14}	C_{13}	C_{12}	0	0
Full-field data word 0 crc<5:0>	$\bar{P}$	P	C_5	C_4	C_3	C_2	C_1	C_0	0	0
Full-field data word 1 crc<11:6>	$\bar{P}$	P	C_{11}	C_{10}	C_9	C_8	C_7	C_6	0	0
Full-field data word 2 crc<15:12>	$\bar{P}$	P	V	0	C_{15}	C_{14}	C_{13}	C_{12}	0	0
Auxiliary data error flags	$\bar{P}$	P	0	ues	ida	idh	eda	edh	0	0
Active picture error flags	$\bar{P}$	P	0	ues	ida	idh	eda	edh	0	0
Full-field error flags	$\bar{P}$	P	0	ues	ida	idh	eda	edh	0	0
Reserved words (7 total)	1	0	0	0	0	0	0	0	0	0
Checksum	$\overline{S8}$	S8	S7	S6	S5	S4	S3	S2	S1	S0

Error flags

All error flags indicate only the status of the previous field; that is, each flag is set or cleared on a field-by-field basis. A logical 1 is the set state and a logical 0 s the unset state. The flags are defined as follows:

edh – error detected here: Signifies that a serial transmission data error was detected. In the case of ancillary data, this means that one or more ANC data blocks did not match its checksum.

eda – error detected already: Signifies that a serial transmission data error has been detected somewhere upstream. If device B receives a signal from device A and device A has set the edh flag, when device B retransmits the data to device C, the eda flag will be set and the edg flag will be unset if there is no further error in the data.

idh – internal error detected here: Signifies that a hardware error unrelated to serial transmission has been detected within a device. This is provided specifically for devices which have internal data error checking facilities, as an error reporting mechanism.

ida – internal error detected already: Signifies that an idh flag was received and there was a hardware device failure somewhere upstream.

ues – unknown error status: Signifies that a serial signal was received from equipment not supporting this error-detection mechanism.

Checkword values

Each checkword value consists of 16 bits of data calculated using the CRC-CCITT polynomial generation method. The equation and a conceptual logic diagram are shown below:

Checkword (16-bit) = $x^{16} + x^{12} + x^5 + 1$

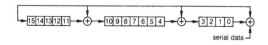

Figure 7.48 The contents of the EDH packet which is inserted in ancillary data space after the associated field.

if active picture is received without error but full field gives an error, the error must be outside the picture. It is then possible to send, for example, pathological test patterns during the vertical interval to stress the transmission system more than regular data to check the performance margin of the system. This can be done alongside the picture information without causing any problems.

In a large system, if every SDI link is equipped with EDH, it is possible for automatic error location to be performed. Each EDH-equipped receiver is connected to a monitoring system that can graphically display on a map of the system the location of any transmission errors. If a suitable logging system is used, it is not necessary for the display to be in the same place as the equipment. In the event of an error condition, the logging system can communicate with the display by dialup modem or dedicated line over any distance. Logging allows infrequent errors to be counted. Any increase in error rate indicates a potential failure that can be rectified before it becomes serious.

An increasing amount of new equipment is available with EDH circuitry. However, older equipment can still be incorporated into EDH systems by connecting it in series with proprietary EDH insertion and checking modules.

References

1. CCIR Recommendation 601-1, Encoding Parameters for Digital Television for Studios
2. Watkinson, J., *Convergence in Broadcast and Communications Media*, Chapter 7. Oxford: Focal Press ISBN 0 240 51509 9 (2001)
3. SMPTE 296M 1280×720 Progressive Image Sample Structure – Analog and Digital Representation and Analog Interface (2001)
4. SMPTE 125M, Television – Bit Parallel Digital Interface – Component Video Signal 4:2:2
5. EBU Doc. Tech. 3246
6. SMPTE 274M Proposed Standard – 1920×1080 Image Sample Structure Digital Representation and Digital Timing Reference Sequences for Multiple Picture Rates
7. CCIR Recommendation 656
8. SMPTE Proposed Standard – Bit Parallel Digital Interface for 625/50 System PAL Composite Digital Video Signal
9. SMPTE 244M, Television – System M/NTSC Composite Video Signals – Bit-Parallel Digital Interface
10. SMPTE 259M – 10-bit 4:2:2 Component and $4F_{sc}$ NTSC Composite Digital Signals – Serial Digital Interface
11. SMPTE 292M – Bit-Serial Digital Interface for High Definition Television Systems
12. Wilkinson, J.H., Digital audio in the digital video studio – a disappearing act? Presented at *9th International AES Conference, Detroit*, Audio Engineering Society (1991)
13. Elkind, R. and Fibush, D., Proposal for error detection and handling in studio equipment. Presented at *25th SMPTE Television Conference, Detroit* (1991)
14. SMPTE RP165 – Error Detection Checkwords and Status Flags for use in Bit-Serial Digital Interfaces for Television

8

Practical video interfacing

Having covered a lot of theoretical ground and described the detail of a number of standards earlier in this book, the purpose of this chapter is to give practical advice in configuring, timing, testing and maintaining digital video interface systems. The techniques and tools needed to test digital interfaces are quite different to the traditional analog approach and will be described here. The great similarity between HD-SDI and SD-SDI means that the same installation and test principles apply. In practice the higher frequency of the HD signal means that the effect of a shortcoming will be more serious in HD.

8.1 Digital video routing

Digital routers have the advantage that they need cause no loss of signal quality as they simply pass on a series of numbers. Analog routers inevitably suffer from crosstalk and noise, however well made, and this reduces signal quality on every pass.

The parallel router is complex because of the large number of conductors to be switched and is obsolete. A serial router is potentially very inexpensive as it is a single-pole device. It can be easier to build than an analog router because the digital signal is more resistant to crosstalk. Large routers can easily be implemented with SDI and the long cable drive capability means that large broadcast installations can be tackled.

A serial router can be made using wideband analog switches, so that the input waveform is passed from input to output. This is an inferior approach, as the total length of cable that can be used is restricted; the input cable is effectively in series with the output cable and the analog losses in both will add.

The correct approach is for each router input to reclock and slice the waveform back to binary. Figure 8.1 shows a typical reclocking SDI router. Following the input reclocking stage the actual routing takes place on logic level signals prior to a line driver that relaunches a clean signal. The cables to and from the router can be maximum length as the router is effectively a repeater.

It is not necessary to unscramble the serial signal at the router. A phase-locked loop is used to regenerate the bit clock. This rejects jitter on the incoming waveform. The waveform is sliced, and the slicer output is sampled by locally regenerated clock. The result is a clean binary waveform, identical to the original driver waveform. The Gennum GS9005 is an equalizer/reclocker suitable for such an application because it does not descramble the data stream.

The router is simply a binary bitstream switch, and is not unduly concerned with the meaning or content of the bitstream. It does not matter whether the bitstream is HD, PAL, NTSC or 4:2:2 or whether or not ancillary data are carried – the information just passes through.

The only parameter of any consequence is the bit rate. HD runs at 1.485 Gbits/s, SD component runs at 270 or 360 Mbits/s, PAL runs at 177 Mbits/s and NTSC runs at 143 Mbits/s. Whilst it is possible to make

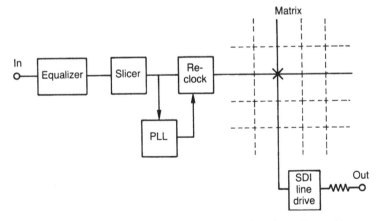

Figure 8.1 A router for SDI signals should be constructed as shown here with a reclocker/slicer at each input to restore the received waveform to clean binary logic levels. The routing matrix proper is then a logic element which is followed by SDI line drivers. If this is not done the output cable is electrically in series with the input cable and the performance margin will be impaired. There is no need to descramble or decode the signal as the router is not interested in its meaning.

a router that will handle HD or SD, this would currently be an expensive option. Many multi-standard routers require a link or DIP switch to be set in each input in order to select the appropriate VCO centre frequency. A separate VCO adjustment may be present for each standard. Otherwise units can be standards independent, which allows more flexibility and economy. With a mixed standard router, it is only necessary to constrain the control software so that inputs of a given standard can only be routed to outputs connected to devices of the same standard and one router can then handle component and composite signals simultaneously.

8.2   Timing in digital installations

The issue of signal timing has always been critical in analog video, but the adoption of digital routing relaxes the requirements considerably. Analog vision mixers need to be fed by equal length cables from the router to prevent propagation delay variation. In the digital domain this is no longer an issue as delay is easily obtained and each input of a digital vision mixer can have its own local timebase corrector. Provided signals are received having timing within the window of the inputs, all inputs are retimed to the same phase within the mixer.

Figure 8.2 shows how a mixing suite can be timed to a large SDI router. Signals to the router are phased so that the router output is aligned to station reference within a microsecond or so. The delay in the router may vary with its configuration but only by a few microseconds. The mixer reference is set with respect to station reference so that local signals arrive towards the beginning of the input windows and signals from the router (which, having come farther, will be the latest) arrive towards the end of the windows. Thus all sources can be retimed within the mixer and any signal can be mixed with any other. Clearly the mixer introduces delay, and the signal feedback to the router experiences further delay. In order to send the mix back to the router a frame synchronizer is needed on the output of the suite. This introduces somewhat less than a frame of delay so that by the time the signal has re-emerged from the router it is aligned to station reference once more, but a frame late. An installation of this kind relies on a gen-lockable sync pulse generator having multiple outputs with independent phase control.

In an ideal world, every piece of hardware in the station would have component SDI outputs and inputs, and everything would be connected by SDI cable. In practice, unless the building is new, this is unlikely. However, there are ways in which SDI can be phased in alongside analog

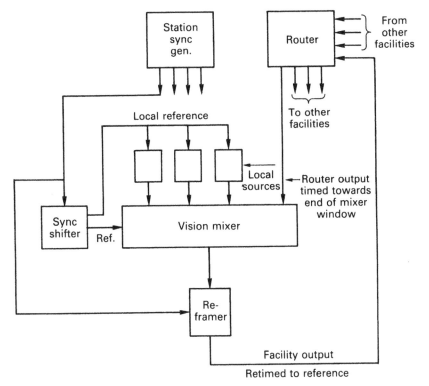

Figure 8.2 In large systems some care is necessary with signal timing as is shown here. See text for details.

systems. An expensive way of doing this is to fit every composite device with coders and decoders, and every analog device with convertors so that a component SDI router can be used alone. Figure 8.3 shows an alternative. The SDI router is connected to every piece of digital equipment. Analog equipment continues to be connected to an analog router, and interconnection paths, called gateways, are created between the two routers. These gateways require convertors in each direction, but the number of convertors is much less than if every analog device was equipped. The number of gateways will be determined by the number of simultaneous transactions between analog and digital domains.

In many cases the two routers can be made to appear like one large router if appropriate software is available in a common control system. This will only be possible if the SDI router is purchased from the same manufacturer as the existing analog router or if the SDI router manufacturer offers custom software.

Whilst modern DVTRs incorporate the audio data in the SDI signal, adapting older devices to do this and subsequently demultiplexing the

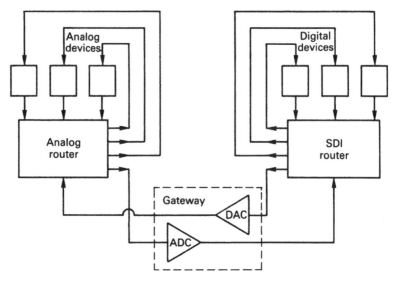

Figure 8.3 Digital routers can operate alongside existing analog routers as an economical means of introducing digital routing.

audio could prove expensive. DVTRs tend also to be inflexible in their audio selection. For example, it may not be possible to record certain audio channels from embedded SDI data at the same time as other channels from the AES/EBU inputs. Also the embedded audio may not contain timecode even though the standard allows it. As a result in some installations it may be more appropriate to retain an earlier audio router, or to have a separate AES/EBU digital audio router layer controlled in parallel with the SDI router.

If the analog router handles composite signals, then coders and decoders will be needed in addition to the convertors. In this case the use of gateways obviates unnecessary codecs when analog devices are connected together.

8.3 Configuring SDI links

The viability of an SDI link is governed primarily by data rate and proposed distance. It is quite easy to establish what grade of cable will be required from the tables in Chapter 7. As can be seen from Figure 8.4, the data integrity deteriorates rapidly beyond a critical cable length. The sudden upswing in bit error rate is known as the 'crash knee' and for reliability only operation to the left of the knee is possible. Figure 8.4 shows

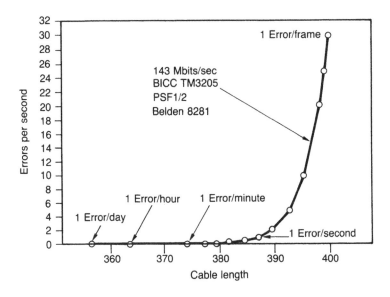

Figure 8.4 Because of the multiplicative effect of the large number of factors causing signal degradation the error rate increases steeply after a certain cable length. This sudden onset of errors is referred to as the 'crash knee'.

Bit error rates (BER)	NTSC	PAL	270 Mb	360 Mb
1 bit error/field	4.2×10^{-7}	2.8×10^{-7}	1.8×10^{-7}	1.3×10^{-7}
1 bit error/second	7.0×10^{-9}	4.7×10^{-9}	3.1×10^{-9}	2.3×10^{-9}
1 bit error/minute	1.2×10^{-10}	7.8×10^{-11}	5.1×10^{-11}	3.8×10^{-11}
1 bit error/hour	1.9×10^{-12}	1.3×10^{-12}	9.0×10^{-13}	6.4×10^{-13}
1 bit error/day	8.1×10^{-14}	5.4×10^{-14}	3.5×10^{-14}	2.7×10^{-14}
1 bit error/month	2.6×10^{-15}	1.8×10^{-15}	1.2×10^{-15}	8.9×10^{-16}
1 bit error/year	2.2×10^{-16}	1.5×10^{-16}	1.0×10^{-16}	7.4×10^{-17}

the situation for SD. For HD, a similar trend will be observed, but the cable length values should be divided by approximately two.

The existence of the crash knee makes it obvious that quoted cable length figures should not be stretched. As SDI is a point-to-point interface all receivers are equipped with an internal terminator. Thus there is no such thing as passive loop-through and use of a coaxial T-piece for monitoring is ruled out. Active loop-through means that, in the case of a power failure, the loop-through signal will fail. If it is required to drive multiple destinations, a digital distribution amplifier will be needed.

If the distance required is excessive even for the best grade of cable then a repeater will be required. Unlike an analog repeater, a properly engineered digital repeater causes no generation loss and minimal delay. Figure 8.5 shows a typical reclocking repeater. The repeater is not interested in the

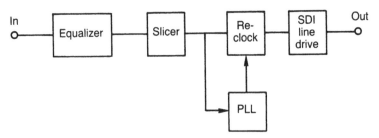

Figure 8.5 In a reclocking repeater the input signal is equalized, sliced and reclocked with a phase-locked loop to recover a clean binary signal which then passes to an SDI line driver.

meaning of the data and need not descramble or deserialize the data stream. It is only necessary to reclock with a phase-locked loop to reject jitter and slice to reject noise. The resultant clean logic level signal can then be supplied to a further SDI cable driver.

Such a simple reclocking repeater is of limited use in an EDH system as there is no ability to distinguish errors occurring before or after the repeater. If this is important an EDH supporting repeater will be necessary. By definition such a device must descramble and deserialize the data stream in order to make the error-detecting checks on the incoming data. As a result it will cause a greater signal delay than a simple reclocking repeater.

8.4 Testing digital video interfaces

Once video and audio are converted to the digital domain, they become data, or numbers, and if those numbers can be delivered to the other end of a digital interface unchanged then the interface has not caused any loss of quality. This is one of the strengths of digital technology. In the absence of data reduction techniques, the quality is determined in the conversion process and can then be maintained in transmission and recording. In contrast analog signals are subject to generation loss in every recording and to noise and distortion in every transmission. This analog heritage has led to a philosophy where the analog waveform is monitored at every stage so that some adjustment can be made to minimize the quality loss. The waveform monitor and vectorscope tradition is so strong that despite the transition to the radically different digital technology many people think no new monitoring methods are needed.

Unfortunately, traditional analog testing techniques reveal nothing about a digital interface or recorder. Consider the system of Figure 8.6, which could be composite or component. An ADC converts the input

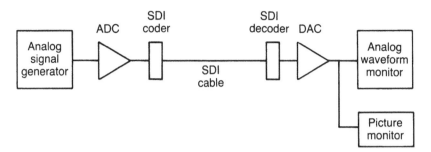

Figure 8.6 Testing waveforms in the analog domain as shown here reveals nothing about the performance margin of the digital interface.

waveform to data that are transmitted by the interface. A DAC converts the received data to analog video once more. If a waveform monitor and vectorscope are connected to the DAC, what information is revealed about the interface? If it is assumed that the interface is not suffering bit errors, the monitoring tells us how good the ADC and DAC are, but reveals nothing about the performance of the interface. The interface could be working with 20 dB of noise immunity, or it could be within a whisker of failure. Should the system be marginal such that one bit fails per minute, this will be invisible on an analog monitoring system in the presence of programme material and may just be detectable on colour bars. If the problem is due to a phase-locked loop drifting in an SDI receiver or damp penetration in a cable, it is going to get worse and in the absence of a warning the result will be a sudden failure.

Three distinct testing areas are required in digital video systems. First, on installation, it should be possible to verify that the link is working with an adequate safety margin and that the length of the link is not excessive for the cable type selected. Second, it is necessary to test the data integrity of the link to ensure that the BER (bit error rate) is acceptable and remains acceptable when the system is stressed or *margined* beyond the conditions it will experience in service. Third, digital systems can suffer from a problem having no parallel in the analog domain. This is the protocol error where the data transmission is flawless but the two units concerned cannot understand each other.

Although the SDI signal is digital in that it carries discrete data, it is an analog waveform as far as the cable and receiver equalizer are concerned. Waveform distortions will occur in the cable and noise and jitter will be added. The magnitude of these distortions indicates the likely reliability of the channel. Figure 8.7 shows that a correctly functioning digital receiver is specifically designed to reject the analog waveform distortions by making discrete decisions. By definition, in doing so it denies us knowledge of the signal quality. Thus what is needed is a complementary

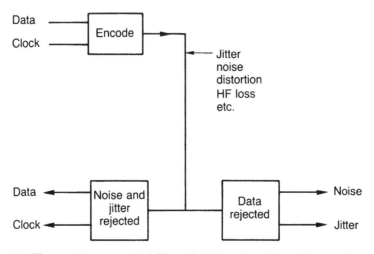

Figure 8.7 The reception process is deliberately designed to reject analog waveform distortions whereas only by measuring these can signal integrity be assessed. Thus the measurement process is diametrically opposed to normal reception and needs special techniques.

approach. Instead of rejecting the distortions to obtain discrete data, what is needed is a system that rejects the data in order to measure the magnitude of the distortions.

One approach is to assess the eye pattern generated by the received waveform. After equalization the eye opening should be clearly visible, and the size of the opening should be consistent with the length and type of cable used. If it is not, the noise and/or jitter margin may be inadequate. However, testing the eye pattern on the SDI requires a fast oscilloscope. Some test equipment samples the SDI waveform at high speed and buffers it so it can be seen on a conventional display. Many SDI receivers have a dedicated, buffered, test point for eye pattern monitoring.

Inspection of the eye pattern is acceptable for establishing that the basic installation is sound and has proper signal levels, termination impedance and equalization, but is not very good at detecting infrequent impulsive noise. Contact noise from electrical power installations such as air conditioners is unlikely to exist for long enough to find an eye pattern display. The technique of signature analysis is better suited to impulsive noise problems.

8.5 Signature analysis

In a digital routing system, reliability is synonymous with data integrity. A data integrity testing system doesn't care what the video waveform is, or indeed that it is a video waveform at all. A data integrity system considers the digital TV field as a block of binary data and simply checks whether that data was received with bit accuracy or not. The message is to forget the pictures and worry about the data.

Signature analysis is a data integrity testing technique using large quantities of data and tests down to extremely low bit error rates. As a digital interface having no bit errors is transparent, signature analysis is a useful way of verifying the transparency of a channel following installation or maintenance. Signature analysis requires a stationary test signal to be applied at the transmitting end of the interface. Stationary means that in the case of component video every frame contains identical data. In the case of composite video the data repeats every four fields (NTSC) or eight fields (PAL). Digitally generated colour bars are suitable but other patterns work equally well. The received data are then processed to generate a value known as a signature. In typical equipment the signature will be a four-digit hexadecimal number. If the transmitted data are always the same, the received signature should always be the same. Any change in the signature indicates that an error has occurred.

The signature generation process divides the incoming bitstream by a polynomial, as was described in Chapter 3. The remainder of the division expresses an entire frame in component or an entire colour sequence in composite as a single signature word. A change of a single bit is enough to change the signature. Any number of bit errors up to the number of bits in the word will guarantee to change the signature, whereas larger numbers of errors are detected with a high degree of probability. In the presence of high error rates the occasional misdetection is irrelevant as the goal of the testing is to determine whether or not remedial action is necessary.

Signature analysis is similar in principle to EDH except that the signature is not transmitted with the data. This avoids complexity at the transmitting end and the need to reserve word positions in the data. As the transmitted data is not a codeword, the remainder will not necessarily be non-zero but could have any value. An error is indicated when the received signature changes, not by its absolute value.

Signature analysis can be used in two ways shown in Figure 8.8. In absolute analysis, shown in (a), the test signal comes from a special generator that displays the signature of the data. The data are fed down the channel under test and then into the signature analyser. The signature displayed on the analyser is compared with the signature on the generator. If the two remain identical for an extended period, then no errors are occurring.

It is also possible to use a relative detection method, shown in (b). In relative signature analysis, any stationary generator can be used. The correct signature is not known, but if errors occur, the displayed signature will not

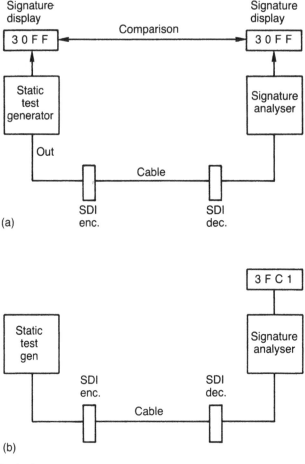

Figure 8.8 In absolute signature analysis (a) the signature at the generator is known and can be compared with the received signature. In relative analysis (b) the signature is unknown but if the generator is known to be static the signature should be constant. Thus changes in the received signature indicate errors.

be stable but will change. Thus in relative signature analysis the goal is not that the signature is correct but that it should not change for an extended period. Relative signature analysis cannot detect permanent errors, such as a stuck bit in a parallel interface, as the same signature will always be obtained and so its use is restricted to systems having no hard faults.

In case of doubt, the test pattern signature can easily be obtained by connecting the generator directly to the signature analyser as well as to the path under test.

Signature analysers can be designed to work on specific parts of the transmission only. If the interface is carrying programme material, the signature will vary from frame to frame. However, the ancillary data slots can still be used for signature analysis.

In component systems, the signature analyser may be set to operate on only one selected component. Some machines can be set to operate only on selected bits in the sample, making stuck bits in parallel systems very easy to find.

As signature analysis works in the data domain, it cannot be used to test a channel in which the received data are not necessarily the same as the transmitted data. There are a number of cases in which this could occur. If the digital signal is returned to the analog domain and then converted back to digital, noise in the analog domain will cause data differences. In DVTRs, uncorrectable errors due to dropouts result in concealments and changed data values. Thus signature analysis can be used to detect concealment. Note that concealment errors are much less visible than general transmission errors by a factor of about 1000:1, i.e. 1000 concealment errors are about the same level of visibility as one transmission error.

Systems using compression are not transparent and signature analysis is of no use if a compression codec is included in the test channel. In order to test the channel the generator must be connected *after* the compressor and *before* the decoder.

8.6 Margining

It is a characteristic of digital systems that failure is sudden because deteriorations in the signal are initially rejected until they grow serious enough to corrupt data. Put simply, just because a digital system is working today, there is no guarantee that it will work tomorrow unless its performance margin can be measured. Margining is a long established technique used in the computer industry to prove the reliability of a digital process by testing it under conditions more stressful than will be encountered in service. The degree of additional stress that can be applied before failure is a measure of the performance margin. As digital video is only data, margining techniques can be applied to it with great success and if used correctly will give a much needed confidence factor.

There are two ways in which an SDI system can be stressed. The first of these is to use a special signal generator producing pathological test patterns. These are bitstreams that mimic the convolution process of the SDI scrambler and result in channel signals containing less clock content and lower frequencies than usual. Receivers find it harder to decode pathological signals and so they will result in errors unless the system is in good shape.

A simple alternative to the pathological test signal is the use of a cable simulator that has the same effect as increasing the length of the cable between transmitter and receiver. The Faraday 'cable clone' is the best known of these devices. Figure 8.4 shows the non-linear relationship between cable length and error rate and illustrates the 'crash knee' in the characteristic. Testing with a cable simulator depends upon the existence of the crash knee. Figure 8.9(a) shows how the test is made. The cable

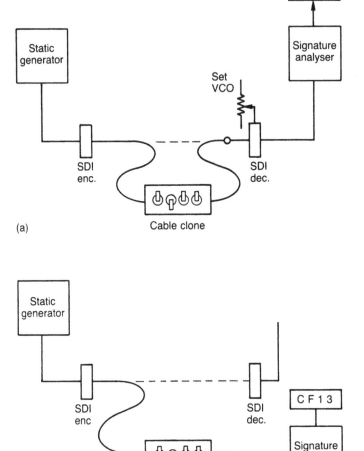

Figure 8.9 Margin testing requires the cable under test to be broken and a 'cable clone' or simulator to be inserted as in (a). The cable length is artificially increased until the crash knee is reached. The configuration in (b) is incorrect as the receiver to be used in service is not being tested. In fact the signature analyser's receiver is being tested: a meaningless exercise.

under test is temporarily broken and the 'cable clone' is inserted. An error-monitoring system such as EDH or a signature analyser is connected after the receiver under test.

The configuration shown in (b) is incorrect as it only tests the cable and the transmitter in conjunction with the receiver in the signature analyser. The important receiver is not tested. The cable clone must be installed in the cable and the signature analyser must be connected after the receiver under test.

With the cable clone set to bypass, the system should show no errors. If errors are detected the fault should be rectified. Starting with an error-free system the cable length is gradually increased until a rapid increase in error rate indicates that the crash knee has been reached. The additional length of cable needed to reach the crash knee is a direct measure of the performance margin or head height. It will be seen from Figure 8.4 that the error rate changes from negligible to intolerable with an increase in length of only 20–30 metres. Clearly if less than this figure is achieved in the margining test the system is marginal and should not be put into service.

One potential problem area frequently overlooked is to ensure that the VCO in the receiving phase-locked loop is correctly centred. If it is not, it will be running with a static phase error and will not sample the received waveform at the centre of the eyes. The sampled bits will be more prone to noise and jitter errors. VCO centring can be checked in a number of ways. If a frequency meter is available, this can be used to display the VCO frequency without input. Another method is to display the control voltage. This should not change significantly when the input is momentarily disconnected. However, the best method is to adjust for minimum error rate in conjunction with margining. Errors can be seen on a video monitor.

Using a cable clone, cable length is added until a slight error rate is caused. The VCO centre frequency is now adjusted one way or the other to see if the error rate can be reduced or even eliminated. If there is a range of error-free adjustment, more cable length should be switched in to make a finer adjustment.

In many receivers, the phase-locked loop will stay in lock more readily than it can achieve lock. Thus the VCO can be misadjusted whilst a signal is being received and will continue to lock to it. However, if the input signal is interrupted, lock will be lost. Consequently a good method is to adjust for the minimum errors after an interruption of the signal.

Some early SDI receivers run quite hot and the VCO centre frequency changes with temperature. Placing cards on an extender board may change the airflow enough to alter the centre frequency.

8.7 Protocol testing

In digital systems it is possible for problems to occur in which the data transmission is flawless but communication is still not achieved. This can happen where the transmitter and receiver have incompatible protocols. The protocol of a signal includes the nature and positioning of TRS patterns, the location of EDH blocks and embedded audio. A further consideration is that in order correctly to adjust analog-to-digital convertors it is necessary to monitor the actual code values coming from a convertor and compare them with the original analog voltages.

Such problems can only be addressed by use of a logic analyser. General-purpose logic analysers can be used on parallel interfaces, but for best results a dedicated digital video analyser is to be preferred. Figure 8.10 shows the layout of a video analyser. An SDI receiver and/or a parallel receiver drive a decoder that identifies TRS patterns for synchronizing purposes. Incoming words are written into a page of RAM that can hold the contents of a few lines of video. The analyser thus takes a snapshot of interface activity. The point within the frame at which the snapshot is taken is determined by the trigger criteria provided to the RAM write logic.

Once the RAM is written in real time with the digital video snapshot, the data are frozen and can be inspected using the associated PC and its display. Signals can be displayed graphically as they would appear on conventional analog waveform monitors or vectorscopes, but can also be displayed as tabular data so that the exact binary word values and positions can be

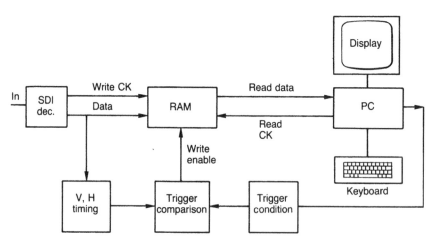

Figure 8.10 A video analyser allows a snapshot of digital video interface activity to be captured in RAM. This data can then be inspected at leisure using a small computer.

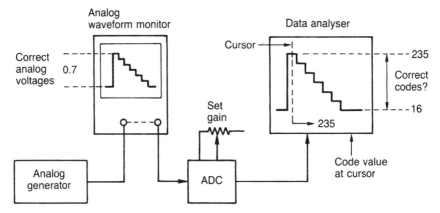

Figure 8.11 Using a video analyser to set the gain of an ADC. The ADC output data are compared with standard values on a known part of the analog waveform.

established. Any departures from correct protocol can be detected by comparing the data with the relevant standards.

Figure 8.11 shows how a video analyser is used to set up a component ADC. Convertor set-up is critical as digital systems keep the data the same all down the line, so it has to be right from the outset. A precise analog test pattern generator is used to produce colour bars. The signal is looped through an analog waveform monitor to the ADC where it should be terminated with the actual terminator to be used in practice. This is necessary because terminator tolerance is such that changing a terminator can change the analog level by several digital code values. The generator level is adjusted until the waveform monitor displays exactly the correct amplitude in, say, the white bar.

The analyser then captures the ADC output and the black bar is located in waveform mode with the cursor. The actual data values at the cursor position are displayed, and these can be compared with the ideal so that any offset in the convertor can be revealed and adjusted out. By selecting the white bar the luminance convertor gain can be adjusted. Some analysers allow timed retriggering so that a new snapshot is taken periodically. This mode is useful when making dynamic gain adjustments. The colour difference signal gains are set up in a similar manner.

The analog Y/C timing should be checked with a suitable signal such as bowtie before returning to colour bars to check the relative timings of the green/magenta transitions in the three digital components. This is not as easy as it sounds as the colour difference sample spacing is twice that of luminance and the sample values need to be interpreted carefully to find the transition.

Index